KU-283-279

COMMANDOS

Investigative and Military books by John Parker:

The Gurkhas: The Inside Story of the World's Most Feared Soldiers
SBS: The Inside Story of the Special Boat Service
Death of a Hero: Captain Robert Nairac, GC, and the Undercover War in Northern Ireland
The Killing Factory: The Secrets of Germ and Chemical Warfare
The Walking Dead: Judge Liliana Ferraro, Mafia Hunter
At the Heart of Darkness: The Myths and Truths of Satanic Ritual Abuse
King of Fools: The Duke of Windsor and his Fascist Friends

COMMANDOS

The Inside Story of Britain's Most Elite Fighting Force

John Parker

HEADLINE

First published in 2000
by HEADLINE BOOK PUBLISHING

John Parker would be happy to hear from readers
with their comments on the book at the following
email address: johnparker@wyndham.freeserve.co.uk

10 9 8 7 6 5 4 3 2 1

British Library Cataloguing in Publication Data.
A Catalogue record of this book is available from
the British Library.

ISBN 0 7472 7008 2

Typeset by
Letterpart Limited, Reigate, Surrey

Printed and bound in Great Britain by
Mackays of Chatham plc, Chatham, Kent

HEADLINE BOOK PUBLISHING
A division of Hodder Headline
338 Euston Road
London NW1 3BH

www.headline.co.uk
www.hodderheadline.com

CONTENTS

Acknowledgements		vii
Prologue		1
1	Bad Beginnings	5
2	The First Commandos	17
3	Better Results	35
4	Not According to Plan, Again	43
5	The Exploding Trojan Horse	61
6	A Cover-up	77
7	Dieppe: the Massacre	89
8	Hitler's Revenge	107
9	The Duke of Wellington's Last Stand	125
10	Send for the Marines? Not Bloody Likely!	141
11	D-Day: Death and Destruction	153
12	Last-chance Saloons	167
13	Here We Go Again!	183
14	The Horror of Korea	197
15	The Defector	209
16	Dear Mum and Dad . . .	225
17	Going South	239
18	Cutting Edge	257
Epilogue: Modern Times		269
Index		273

Acknowledgements

Although originally designated as potential formations under the auspices of the Royal Marines, Commandos came into being during the early stages of the Second World War as army units. Later in the war, the Royal Marines themselves provided Commando formations, and later still the whole Commando organisation was transferred to RM control. It was a contentious issue then, and remained so for many years, with the original army personnel adamantly and sometimes vehemently reminding the world that they formed as volunteers, not as conscripts or as regulars. They were indeed 'irregular' in virtually every respect. Those old divisions began to heal as the two sets of Commandos, army and RM, were thrown together, fighting alongside each other – although they remained quite separate formations. The divisions were highlighted again, however, when the army units were disbanded after the war, leaving the Royal Marines to carry the banner forward. Thus, in these pages I have attempted to reunite the Commandos, bringing to life some of the exploits of the earliest Commando units and onwards to modern times. To achieve this, I have once again relied heavily on personal accounts that relate to virtually every major campaign involving Commandos, whether army or Royal Marines. The words are those of the men themselves, in the 'I was there' mode, often involving harrowing and dramatic events, evidently ingrained on the memory to be recalled in minute detail years later. The author's own contribution was to piece these accounts together and add the side action, the surrounding history and observations to these events.

Once again, I am deeply indebted to the large number of people who have assisted with this project, and especially to those who contributed their time and effort in the process of bringing this story to life. The result is an account largely told by the men themselves, with the recollections obtained through tape-recorded interviews conducted either personally with myself, my colleague Alastair McQueen or recorded for posterity at the Sound Archive of the Imperial War Museum. Other material was obtained from transcripts, diaries and original War Office and Ministry of Defence documentation researched at the Imperial War Museum, the Royal Marines Museum at Southsea and the Public Records Office at Kew. I am also indebted to the staff at the IWM Sound Archive for their tremendous cooperation in assisting in the searching out of specific tapes relating to particular events, and to Matthew Little at the RM Museum for his help in both documentary and photographic research.

John Parker
May 2000

Prologue

Hitler was rattled. His armies had invaded, crushed and occupied virtually the whole of Continental Europe, forced the ignominious retreat of the British Expeditionary Force from Dunkirk, battened down Scandinavia, opened up the Russian front and taken pole position in North Africa. The United Kingdom, alone in the struggle until the Japanese bombed Pearl Harbor, had been niggling away around the edges of the Nazi Fortress Europe and dashing menacingly behind the lines of Rommel's Afrika Korps for months, causing enough trouble for the German leader to issue an infamous edict.

The content went against all the recognised conventions of war and was based on the premise that British Special Forces, known collectively at that time as Commandos, were nothing short of a bunch of murderous thugs and cutthroats who were ignoring the rules, wantonly and often clandestinely killing German soldiers and civilians by every means possible, effectively refusing to take prisoners by shooting them before they had the chance to put their hands up. The fact that Winston Churchill himself had called for the creation of such a force to perform raids deep in enemy-held territory 'leaving a trail of German corpses behind them' may well have contributed to the Germans' claim, which was not altogether without foundation.

Hitler's instruction took the form of a top-secret memo to his most senior aides in the German High Command – initially just 12 copies of it were produced, and thereafter the order was to be filtered down the line for immediate action. It has often been overlooked by military historians, lost to some extent under the weight of other, more outrageous Nazi edicts and partly because there was no desire to delve too deeply into the reasons, which will become clear as these pages progress:

Führer Order Concerning
Handling of Commandos
TOP SECRET

The Führer No. 003830/ 42 g. Kdos.
Führer HQ, 18 Oct. 1942
12 copies, 12th copy.

1. For some time our enemies have been using in their warfare methods which are outside the international Geneva Conventions. Especially brutal and treacherous is the behaviour of the so-called Commandos, who, as is established, are partially recruited even from freed criminals in enemy countries. From captured orders it is divulged that they are directed not only to shackle prisoners, but also to kill defenceless prisoners on the spot at the moment in which they believe that the latter as prisoners represent a burden in the further pursuit of their purposes or could otherwise be a hindrance. Finally, orders have been found in which the killing of prisoners has been demanded in principle.

2. For this reason it was already announced in an addendum to the Armed Forces report of 7 October 1942 that in the future Germany, in the face of these sabotage troops of the British and their accomplices, will resort to the same procedure, i. e. that they will be ruthlessly mowed down by the German troops in combat, wherever they may appear.

3. I therefore order: From now on all enemies on so-called Commando missions in Europe or Africa challenged by German troops, even if they are to all appearances soldiers in uniform or demolition troops, whether armed or unarmed, in battle or in flight, are to be slaughtered to the last man. It does not make any difference whether they are landed from ships and aeroplanes for their actions, or whether they are dropped by parachute. Even if these individuals, when found, should apparently be prepared to give themselves up, no pardon is to be granted them on principle. In each individual case full information is to be sent to the O. K. W. for publication in the Report of the Military Forces.

4. If individual members of such Commandos, such as agents, saboteurs, etc. fall into the hands of the military forces by some other means, through the police in occupied territories for instance, they are to be handed over immediately to the SD. Any imprisonment under military guard, in PW stockades for instance, etc. is strictly prohibited, even if this is only intended for a short time.

5. This order does not apply to the treatment of any enemy soldiers who, in the course of normal hostilities (large-scale offensive actions, landing operations and airborne operations), are captured in open battle or give themselves up. Nor does this order apply to enemy soldiers falling into our hands after battles at sea, or enemy soldiers trying to save their lives by parachute after battles.

6. I will hold responsible under Military Law, for failing to carry out this order, all commanders and officers who either have neglected their duty of instructing the troops about this order, or acted against this order where it was to be executed.

[signed]
Adolf Hitler

A further 22 copies of the edict were passed on to senior army officers by Hitler's Chief of Staff, Army:

Headquarters of Army,	Secret
No. 551781/42g.k.: Chefs W.F St/Qy	F.H. Qy, 19.10.42
	22 Copies
	Copy No. 21

The enclosed Order from the Führer is forwarded in connection with destruction of enemy Terror- and Sabotage-troops. This order is intended for Commanders only and is under no circumstances to fall into Enemy hands. Further distribution by receiving headquarters is to be most strictly limited. The Headquarters mentioned in the Distribution list are responsible that all parts of the Order, or extracts taken from it, which are issued again are again withdrawn and, together with this copy, destroyed.

Chief of Staff of the Army
Signed: JODL

Hitler's order was to be acted on to the letter, and countless British and Allied soldiers captured by Nazi soldiers under circumstances that fell within the broad terms of Hitler's demands were executed, including survivors of the famous Cockleshell Heroes raid on enemy shipping in the Gironde estuary, even though they had already been taken prisoner; numerous such incidents will be recorded in these chapters.

Who were these men who were causing the Nazis so much grief? And were they really the brigands that they were made out to be? Their appearance in the British military goes back to the spring of 1940, when

Norway called for assistance in the face of a German invasion at the same time as tens of thousands of British troops and their hardware were being ferried across the Channel to confront the anticipated German advance into the Low Countries. The Commandos were formed initially as independent companies that were, in reality, small private armies of eager but untrained men, short of weapons, unwanted by the majority of the British army generals and led, by and large, by a colourful collection of military idealists and eccentrics.

Those who witnessed their early training manoeuvres would never have believed they were capable of even a pinprick against the mighty German army. Indeed, from the first-hand accounts of the fiasco-ridden early days of their formation, when green volunteers were thrown immediately into acts of daring in which the odds of safe return were stacked heavily against them, it is a wonder that they did.

So, let us go back to the beginning of their story, to those early months of the conflict when the men they were to call Commandos began their great adventure, volunteering for very active service in units which until that moment did not exist anywhere in the British military establishment.

Chapter One

Bad Beginnings

On the 'needs must' principle, all angles and possibilities were being examined as the United Kingdom stood on the brink of potential catastrophe. Among them were some wild schemes emanating from papers gathering dust in the War Ministry files, written during various studies since the First World War on what were termed 'guerrilla raiders' – small collections of troops who could move quickly, go anywhere, do virtually anything and who were unencumbered by the logistics of large-scale troop movements. These parties, had they ever got off the ground, were originally designated to be run under the auspices of the Royal Marines. As early as 1924, the Madden Committee on defence, which examined some of the military disasters of the First World War – including huge losses incurred in amphibious troop landings at Gallipoli – proposed that a 3,000-strong Royal Marines brigade be set up to undertake raids on the enemy coastlines and bases. But like so many other shortfalls in the British military contingency in 1939, not very much had been done about getting such groups formed.

First, there was a shortage of money for such luxuries and, second, there was considerable resistance to them from those among the top brass who were determined to stick to conventional methods; guerrilla tactics, they stated with forceful rhetoric, had no place in the British armed forces. Thus, guerrilla units in the British army were nowhere in sight when the Second World War approached the start-line, and when the idea was eventually resurrected the Royal Marines were already hard pressed enough and unable take it on; a make-do-or-mend solution was handed to the army and the territorials.

By then Hitler's armies were on the march. He had moved swiftly and ruthlessly to complete his invasion of Poland, which had caused Prime Minister Neville Chamberlain's declaration of war against Germany in

September 1939. A proposed joint British and French offensive to free the Poles and punish Germany was no longer even a pipedream. In any event, the French tactics veered towards the defensive, convinced as they were that the heavily fortified Maginot line would keep the Germans out of France, which, of course, it famously didn't.

After Poland, Hitler ordered his High Command to begin the advance on the Low Countries and France before winter set in, but his generals asked for time to prepare for what they envisaged as an unprecedented blitzkrieg of troops and heavy metal moved by land and air. And so very little happened, and the only sound of military activities outside territory already secured by the Germans was the clump, clump, clump of British soldiers' boots foot-slogging towards a front line across the English Channel that did not, in truth, exist. In fact, there was so little activity in the first months after the declaration of hostilities that the American newspapers dubbed it the Phoney War. The thousands of young men being drawn progressively into military service were either en route to the BEF in France or marching up and down the parade grounds of British military bases and generally preparing the defence of the nation's beaches. They called it the Bore War, and it was only partially enlivened by some side action to the main event which suddenly loomed in Scandinavia.

Taking advantage of the impending conflict in Europe, the Soviet Union invaded Finland in October 1939. Stalin's troops were ill-prepared and badly led because so many experienced officers had been taken out during the political purges. Consequently, although outnumbered by at least five to one, the Finns stood their ground and kept the Red Army at bay well into the new year. The Soviet attack on Finland temporarily diverted the interest of the European contenders towards that region, and specifically to Kiruna in northern Sweden, which was Germany's main supplier of iron ore. In winter, the ore was shipped through the ice-free Norwegian port of Narvik on the western seaboard and then through neutral Norwegian waters to Germany. On the pretext of supporting Finland against the Red Army, the British had, with French acquiescence, worked out that on the way they could occupy Narvik and Kiruna, the two key centres for the shipping of iron ore and halt the flow of vital supplies into Germany.

It would require the cooperation of both Norway and Sweden. Both nations refused, in the hope of maintaining their neutrality, and a battalion of the 5th Battalion Scots Guards, specially trained for ski manoeuvres and already on its way to Finland, had to be brought back. As the weeks wore on, Stalin was making little headway in Finland and, fearing outside intervention, extricated the Soviet Union by agreeing terms with the Finns on 8 March 1940, leaving the British and French to find another pretext for their proposed occupation of Narvik and Kiruna.

This was to be achieved by sending the British navy to distribute a large number of mines just outside Narvik harbour in the hope of provoking German reaction, which would allow the Allies to spring to the aid of Norway and into Narvik.

That didn't work, either. Prevented by bad weather and adverse intelligence reports, the German High Command delayed its invasion of the Low Countries still further and instead focused on the developing situation in Scandinavia. The upshot was Hitler's approval of an incursion into Denmark and Norway on 7 April, the latter through eight ports from Narvik and on down around the western and southern coastline to Oslo and, for the first time ever, using airborne troops to secure inland sites. A British task force, having laid mines around Narvik, was at that moment sailing home and actually passed the German ships without seeing them, leaving the way clear for largely unopposed landings on the morning of 9 April.

As German warships appeared off the Norwegian coast, the airborne invasion began inland. Within forty-eight hours, the Germans had landed seven divisions and captured all the main ports, while the airborne troops secured their position in Oslo and major airports. The weather halted planned parachute drops at Oslo airport, and infantry troops were landed in a succession of Ju-52s to take possession. Five companies of parachute troops did, however, drop at other key airports. The Germans established a firm hold on the southern half of Norway.

Oslo was overrun by noon that day, yet the Norwegian government decided to make a stand and moved to Elverum, there to send word inviting the British and French to dispatch troops immediately to their assistance. The terrain was rugged and difficult, and with more than 350,000 troops already committed to the BEF to confront the German invasion that must surely soon come, the British army agreed at last to the formation of 'guerrilla forces' whose recruitment was based initially on the region in which they lived. The idea was derived from small special military units that first made their appearance among the Boers in the late nineteenth century and which had been the subject of several studies by British military strategists in the intervening years, including the Madden Committee.

The guerrillas were originally used in raids and assaults first against hostile African tribes and then in the Boer War against the British. They were hit-and-run troops who raided installations, sabotaged machinery and communications and attacked enemy troops. The Boers raised them through electoral districts, each area supplying its own force. Because the men had been 'commandeered' by the military, they soon took the name commandos, and self-sufficiency was their trademark: each man was responsible for providing his own horse, they received no pay or uniform and had no permanent headquarters or base. Their tactics centred around lightning strikes on the British forces, and then they melted back into the veldt before the British could react.

The Boer War was extended by at least a year by the activities of the commandos. Afterwards, the nearest the British army came to forming Special Forces units of this kind occurred during the First World War, when T. E. Lawrence led Arab levies in revolt against the Turks and in support of General Allenby's triumphal advance into Damascus in October 1918. In fact, a former pilot who had assisted Lawrence, Major J. C. F. Holland, who had later been badly wounded in Ireland, was given the task of studying guerrilla tactics in the mid-1930s. He was seconded to a research unit in the War Office to lead an appraisal of the Boer commandos, of T. E. Lawrence and of the formation of special groups in more localised wars since 1918, including Britain's campaigns on the North-West Frontier. He was to report to the director of operations at Military Intelligence to a department that became MI(R). Holland was joined by Lieutenant-Colonel C. McV. Gubbins, MC, RA, who had served with him in Ireland as well as in northern Russia in 1919. MI(R) published a series of papers on guerrilla warfare, while Gubbins himself was sent on intelligence-gathering trips to the Danube valley, Poland and the Baltic states. He was in Poland when war broke out and escaped through Bucharest. It was MI(R) which proposed the raising of a battalion trained in the use of skis to be sent to the aid of the Finns in the spring of 1939 before the operation was eventually aborted by the end of the Soviet–Finnish dispute.

The MI(R) unit continued its work on other ideas, notably in the formation of commando-style units for irregular warfare techniques and, after the German invasion of Norway on 9 April 1940, Britain's own and very first Special Forces were finally rushed into service. The move came after a British force under Major-General P. J. Mackesy landed north of Narvik on 14 April. Another force went ashore at Namsos two days later, with more troops on their way, including contingents from the French Chasseurs Alpins, the Foreign Legion and Polish forces who had escaped during the German invasion of their own country. It quickly became apparent, however, that the vast and rugged Norwegian coastline was no place for ill-equipped British troops who had no suitable clothing, boots or transport for such conditions.

General Carton de Wiart took a scratch contingent raised from county regiments, including the Bedfordshire and Hertfordshire. He landed on 14 April with orders to attack German positions at Trondheim, supported by the guns of the Royal Navy ranged along the shore. On that day, HMS *Warspite* sank seven enemy destroyers in the Narvik fjords, while several other enemy ships were scuttled outside Oslo. On land, successes were harder to come by. Most of Wiart's men had hardly ever seen a hill, let alone climbed through treacherous, rocky terrain. Furthermore, while the Germans were kitted-out in white uniforms, were equipped with skis and other made-to-measure accoutrements to suit the conditions, the 'special clothes' provided for British troops consisted of fur coats, thick knitted jumpers, heavy-duty socks and gardening boots that leaked. As the general opined later: 'If my troops wore all of those things together, they were unable to move about and looked like paralysed bears. As far as guns, planes and transport were concerned, I had no worries at all, for no such things were available.'

Wiart was forced to retreat, his force badly mauled and let down.

The British/French expedition to help the Norwegians repel the German invaders was a fiasco from the outset. Troops were needed over a wide area between Narvik in the north and Namsos, 310 miles away, and on towards Oslo, ostensibly to prevent the Germans from setting up air- and sea bases. Step forward the MI(R) unit with its proposals for 'guerrilla companies' to be formed immediately.

They were to be based almost exactly on the Boer commando model, although it has to be said that, brave and keen though they were, the initial line-up did not look at all encouraging. The hurried plan was to form 10 independent companies, each consisting of 21 officers and 268 other ranks, and because time was so short they were to be raised largely from volunteers from the second-line Territorial Army divisions still in the United Kingdom. Ostensibly, each unit would secure men with some form of training in all aspects of military deployment, so that each had its particular specialists in the use of light weaponry, demolition, transport, engineers and medics.

The companies would have no specific barracks or even accommodation provided. They were supervised by a headquarters company manned by officers from the Royal Engineers and the Royal Army Service Corps with an attached intelligence section. Their sole object was to be deployed on raids against enemy positions and, because all those were outside the island waters of the United Kingdom, the companies were to be organised as ship-home units: the ship carrying them to and from operations would be their base, and at all other times they were billeted in private houses in the coastal towns selected by their officers as their home base. Nor were they provided with transport – all movement was achieved by rail.

Since the whole setup was totally new to the British military, there were bound to be teething problems, and there were. Sir Ronald Swayne, who joined No. 9 Independent Company forming at Ross-on-Wye in April 1940, recalled his own experiences in words that speak volumes as to the planning of the raids and the readiness of the volunteers for any scrap, let alone heavy bombardment by enemy guns:

> I was in the 38th Division, which was a Territorial Welsh Border division, and I was selected from my regiment, the Herefordshire Regiment, to take a section of Herefordshire boys, which I did, selecting lads who came really from round my home, actually. It was all very nice and friendly. I called at my home on the way, stayed the night and the following morning my father drove me to Ross. As we approached the place where we were collecting, we saw a figure with his service cap pulled over one eye, with a Royal Welch Fusilier flash, wearing breeches without any stockings, gym shoes and swinging a shillelagh. He was a major, and he had certainly a DCM and bar and, I think, an MM. My father after a few minutes recognised him as a chap named Siddons who'd been his galloper in the Curragh in 1912. He'd been a trooper and then he'd got a commission; he'd been a Black and Tan, and he was a terrifying figure, totally eccentric, usually drunk, anyhow, after about lunchtime. Quite an original person – shall we say? – and we had some wonderful schemes with him when we were training up – if you can call it that – to be sent to Norway to kick the Germans out. He used to lecture us on how to kill people. He had one lecture about killing people with a fork, a tablefork and other instruments in trenches, and that sort of thing. And I was talking to my soldiers afterwards about it, and a nice young farm labourer called Jenkins asked me what I thought of it.
>
> He said, 'I don't think it's right, do you, Mr Swayne?'
>
> I said, 'No. I can't see myself doing it, can you?'
>
> 'No,' he said. 'I think it's very unchristian.'
>
> Well, Major Siddons wasn't a very Christian sort of chap. He was the original model of Brigadier Richie Hook, the fictional character in the trilogy *Men at Arms* by novelist Evelyn Waugh, who was soon to join us in the Commandos. Siddons used to jump up in the rifle butts inviting us to shoot him when we were practising, which was his way of hardening us up, apparently after he had heard this soldier of mine saying that he thought killing people with a tablefork was most unchristian. But we eventually had a lot of characters like that; we did rather collect them. Some of them were quite hopeless because they were too eccentric. And we went off and we were formed into an independent company, which was on

standby to go to Norway. Other companies, formed a week or so before ours, were already on their way, although I have to admit it was all rather amateurish in the beginning. I don't think, at that point, the Jerries would have been too scared of us.

The dash to form the independent raiding companies was on, and by the end of April 1940 the ten units had been formed largely from local TA regiments around the country, from Somerset to East Anglia, the Midlands, the north and on into Scotland and Northern Ireland, each designated simply by the number in the order in which they were formed, and thus became independent companies bearing numbers from one to ten. They were little more than groups of young inexperienced men who a few days before had been engaged in the humdrum life of your average British working-class lad, from farmhands to shopworkers. Among them was Ernest Chappell from Newport, South Wales, who had left school a couple of years earlier at the age of 16 to work in a fish-and-chip shop and had lately been employed in a factory assembling hot-water geysers before joining the Royal Welch Fusiliers:

I had gone through the usual initial training and [in January 1940] we were sent down to the Wiltshire area on some monotonous guard duties at Andover airport and various little Coastal Command radio stations. It was very boring and I asked to be transferred to somewhere where I could see more active service. Eventually, I was posted to No. 9 Independent Company, which was already in training at Ross-on-Wye and headquartered at the drill hall. It was lashing down with rain when I arrived, and people were coming in off an exercise absolutely soaking wet. The commanding officer, Major Siddons, happened to be in the drill hall when a guy came in dripping water and said: 'Sir, please can I be sent back to my unit?'

Siddons, no gentleman, bawled out: 'Sent back to unit . . . why?' And this chappie said: 'All my friends are there.' Siddons looked at him with a stare to kill and said: 'Friends? You haven't got a friend this side of hell, you worm. Go back to your unit and out of my sight.' This was my introduction to the independent companies. Personally, I enjoyed every minute of it. I liked running through the countryside with scruffy clothes on and a bandoleer around my neck, as opposed to the formality of the normal army regime of being smartly turned out, bolt upright and saluting anything that moved. It seemed I was a free agent. We used to wear our kit and a blanket and a mess tin and a bandoleer with ammunition slung around our necks

> – this is the sort of scheme in a kitbag war. We were organised into troops, aimed more at raiding, landing from the sea. We had no special weaponry, nothing elaborate in those early days. We had the old Lee-Enfield rifles, mostly of First World War vintage. I liked the free style of it. Our discipline was self-imposed and brought on by respect for the people we worked with and for. We were relying on one another, and it was made clear what was expected of us. We knew what we were doing, why we were doing something. Whereas in the Royal Welch you were basically told to lie down and if anyone comes over the top, you shoot. But you didn't know why. Whereas with the independent company – certainly later with the Commandos – we were briefed fairly fully on the part we were playing and why we were doing it. It gave you some aim. You knew what you were up to. We had more scope. We operated in smaller numbers, subsections of three or four men. And if we were loose in the woods, we had to make our own minds up about how we handled a situation. We had more freedom of choice. We were left to our own devices. This was the style of our training, although initially there was very little time. We were supposed to land in Norway and operate behind the German lines there within a matter of days.

In fact, there was no real training for the excursion to Norway before the first units were dispatched to Glasgow to await embarkation to join the rest of the British and French troops bound for Narvik to meet the Germans head-on. The lack of preparation verged on the farcical, given the desperately difficult coastline on which they were to land and operate, and the skill of the enemy troops facing them. Many of the men, perhaps the majority, could not swim. Most had never set foot in a boat, some had never been near such an expanse of water or seen such sights as the huge Atlantic waves crashing ashore against the rocks and inlets of the fjords where they were to land. Every step of their journey was a new experience.

There had been no time for organised instruction of any kind, and in any event no facilities or centre for training men for this kind of warfare existed anywhere in Britain; it was, again, a make-do-or-mend situation, as Ronald Swayne remembered: 'The chaps were very eager and keen. They were mostly locally based, so I knew most of them pretty well. My soldiers were Herefordshire boys who came from round me in Herefordshire, country lads basically. And I think they would have been rather good guerrillas eventually, but there was no time for real training. We had to teach most of them to swim to begin with, and we did that by just throwing them in over the side of a whaler.'

Worse still was the shortage of supplies and transport once there, and even getting to their designated locations would prove to be literally a journey into the unknown, given that the only maps available were taken from 'an illustrated guide to the beauty of Norway as a holiday country'. Even so, No. 1 Independent Company was on its way to Norway within ten days of its formation with orders to go ashore at the port of Mo. The only instructions to the company came by way of the Admiralty to the commanding officer:

> Your mobility depends on your requisitioning or commandeering local craft to move your detachments watching possible landing places. Keep attack from the air always in mind. Disperse and conceal but retain power to concentrate rapidly against enemy landing parties. Keep a reserve. Get to know the country intimately. Make use of locals but do not trust too far. Use wits and low cunning. Be always on guard.

Within the week, four other independent companies, Nos. 2, 3, 4 and 5, were dispatched to Norway, each with a given target and under the overall command of Lieutenant-Colonel Colin Gubbins, whose pamphlets on guerrilla raids had set the whole thing moving. He established his headquarters at Bodo, where No. 3 Independent Company was deployed. His orders were: 'Your Companies . . . should not attempt to offer any prolonged resistance but should endeavour to maintain themselves on the flanks of the German forces and continue harrying tactics against their lines of communications.'

To follow such an order under such conditions was well nigh impossible. Confusion reigned, as No. 4 Independent Company, formed at Sizewell, Suffolk, in mid-April, discovered. The company was moved to Glasgow on 5 May, with no firm instructions as to how the men would actually be transported across the North Sea. The company's officers, after a lot of searching around, eventually discovered they were to go aboard the passenger ship *Ulster Prince*, where they would be served with breakfast. Three hours later, the men having been hanging around the dockside, they were finally taken aboard only to discover that the ship had no fresh food, bread or meat. The company CO, Major J. R. Paterson, received some vague orders as they were about to depart, directing the company to land at Mosjøen, acquire some local boats and move west to Sandnessjøen at the mouth of the fjord. Then the orders were changed, and both Nos. 4 and 5 Companies were instructed to disembark at Mosjøen.

They arrived just after midnight on 9 May. They offloaded as much of their stores as they could before the ship sailed again, then trudged to a point five kilometres north of Mosjøen, where the company deployed to platoon positions. Over the course of the next month, the independent companies were moved up and down the western coastline of Norway, engaging the enemy on numerous occasions and generally achieving the harrying tactics for which they had been deployed. On 9 May, too, the second dispatch of troops from the independent companies, the remaining Nos. 6 to 10, were due to gather in Glasgow to prepare for embarkation the following day. Swayne recalled:

> We got into a little passenger ship to go to Norway, sat around for a few hours, sailed up the Clyde and then sailed back again and disembarked. The message came through that our trip had been cancelled. We weren't going to Norway, none of us, none of the remaining companies, and everyone was exceedingly disappointed. Later that night we learned the reason why . . . Hitler had invaded the Low Countries.

As the second wave of the independents forlornly hung around Scottish docksides waiting for further instructions, companies Nos. 1 to 5 were moving into their respective positions to take up the cudgel against the Germans around the Norwegian coast. Even as they did so, the developments on mainland Europe had already set in motion the beginning of a greater calamity for the Allied soldiers of the UK and France that would also seal the fate of the ill-starred effort to save Norway. In the next 24 hours, at dawn on that momentous day, 10 May 1940, German airborne and parachute units piled aboard 42 gliders that were pulled into the air by Ju-52s from Cologne and released into silent flight over Holland and Belgium, disgorging their crack troops to seize vital airports and bridges.

Meanwhile, along a 150-mile front, 28 German divisions were assembled to move into action. The blitzkrieg, launched without warning, came to the Low Countries, and the British Expeditionary Force began its unhappy retreat to the coast. The British nation went into shock, not least among them the military analysts when, in the aftermath, they pieced together the elements of the German invasion strategy: the lightning speed of the panzer divisions that totally overran defences, backed up by the fearsome accuracy of the airborne artillery provided by the Ju-87 Stukas and the devastating – and totally unexpected – arrival of airborne and parachuted troops.

By the end of May the British Expeditionary Force was being evacuated back to England from Dunkirk, followed soon after by the British and French troops when they were pulled out of Norway, leaving that nation also in the hands of the Nazis. Fortunes there, in the end, had been mixed. Though many armchair observers were highly critical of the whole operation, especially around Narvik, some bright spots would have shone even brighter if the troops had not suddenly been withdrawn. The Norwegians, aided by 12,000 British and French, held out in the area between Oslo and Trondheim until 3 May. Around Narvik, 4,600 Germans faced 24,600 British, French and Norwegians backed by the guns of the British navy, which performed some extensive surgery on German shipping. By the end of May, German troops were kicked out of the hills surrounding Narvik and were still retreating when suddenly all changed and the hard-pressed Germans could not believe their luck.

On the morning of 3 June they discovered that the British and French were pulling back then disappearing over the horizon. The orders to withdraw and return to Britain came as the news from France grew more dismal by the second, with capitulation only days away. In the final hours of the evacuation of Allied troops from Narvik, the aircraft carrier HMS *Glorious* and two destroyers were sunk with more than 1,500 men lost. In France, the situation was even worse. The British and French were forced into a narrow beachhead around Dunkirk, and hundreds of small craft of all kinds joined the Royal Navy sealift to rescue 338,226 men from the beaches. In all, 222 naval vessels and 665 other ships were involved in the evacuation. Many thousands were killed, wounded or taken into captivity, and a very large slice of Britain's entire stock of weaponry, tanks, guns, lorries and general stores lay wrecked by the departing troops and burning on the roads and in the fields on every approach to the Dunkirk beaches and other exit points along the coast.

The inventory of weapons and equipment lost was staggering, and it is worth reminding ourselves of the totals: 475 tanks, 38,000 vehicles, 12,000 motorcycles, 8,000 telephones, 1,855 wireless sets, 7,000 tonnes of ammunition, 90,000 rifles, 1,000 heavy guns, 2,000 tractors, 8,000 Bren guns and 400 antitank guns. On 6 June 1940 the War Cabinet was informed that there were now fewer than 600,000 rifles and only 12,000 Bren guns left in the whole of the United Kingdom and the losses would take up to six months to replenish. It was a truly desperate situation that was kept well from public view. George Cook, who we will meet later as a fully fledged Commando, recalled a memorable and oft-repeated scenario that summed up Britain's parlous state:

> We were on aerodrome defence because it was believed that the Germans might land at any moment and send their paras to capture the aerodromes first, as they had done in Belgium and Holland. We dug slit trenches all round the perimeter, and we used to do stand-tos

> at night, a 24-hour guard basically. But we never had any ammunition, none anywhere. I asked one of the officers what we would do if there was hell and the Germans were landing? Did we just shout 'Bang, you're dead' or what? And if they wouldn't drop dead when we shouted, what would happen to us? I was put on a charge for being insolent!

In France, meanwhile, the news lurched from bad to worse: Marshal Henri Philippe Pétain, a First World War hero who had become premier of France the day before, asked for an armistice that was signed on 25 June. Pétain then set up a capital at Vichy in the unoccupied south-east.

Hitler now set his sights on his one remaining active enemy, the United Kingdom . . .

Chapter Two

The First Commandos

Winston Churchill well knew the tale of the Boer Commandos. He had resigned his cavalry commission in 1899 to become a war correspondent and had been captured by the Boers, escaping with a price on his head to return home a national hero. Not surprisingly, therefore, within 24 hours of succeeding Chamberlain as Prime Minister on the day Hitler began his march into the Low Countries, Churchill was looking for proposals for the creation of Special Forces of a kind that did not exist in the British military. He demanded prompt action to raise commando-style seaborne raiding parties, saboteurs, espionage agents and airborne and parachute troops, the latter having so successfully spearheaded the Nazi incursions into Norway, Belgium and Holland. Thereafter and in spite of all the other great worries surrounding him, Churchill took great personal interest in nurturing them into being, forgiving early errors and pointing the way forward to the point of insistence, as will be seen by simply tracking the dates of significant developments in his first month at number 10.

It was apparently clear to him, even then, the way the early part of the war would develop as far as the British were concerned. The British army would need time to recover from Norway and Dunkirk, and in any event the defence of the British Isles was an immediate priority, although largely in the hands of the brilliant and brave Royal Air Force as the Battle of Britain commenced. The whole of Continental Europe was a no-go zone for any form of conventional troop-landings, especially when Italy joined the war as Germany's sidekick. The only viable options for the time being were attacks from the air and clandestine hit-and-run seaborne raids around the coastlines of Nazi- and Italian-held territory, with Gibraltar and Malta providing the key staging-posts to the Mediterranean.

Another reason, too, urged Churchill on to achieve the creation of landing forces of some calibre that could get some form of presence in enemy territory at a time when any mass reinvasion attempts would have been cut to ribbons. It harked back to the First World War and a disaster that no one, especially not Churchill, could forget: Gallipoli, the single most costly adventure in terms of casualties in British military history. The attempts to ferry and land troops into the Dardanelles strait had been devised by Munitions Minister David Lloyd-George, First Lord of the Admiralty Winston Churchill, General Kitchener and Admiral Sackville H. Carden. It was aimed at opening up a new war front to distract the Germans from throwing a new offensive into the stalemate situation in France in the early months of 1915. Lack of planning, bad maps and suicidal landing approaches spelled death to the Allies. The Gallipoli campaign ended with catastrophic casualties: 205,000 of the 410,000 Anglo-Anzac-Gurkha troops called into British service were killed or wounded.

The parallels were striking: for the moment the Germans were back in France and were winning. There was no way that Churchill could land a major force anywhere on Continental Europe, even if he had the troops and equipment to spare – which he didn't. Whether Britain could hold out at all was questionable, and much rested on the ability of the RAF to fight off the nightly attacks by the Luftwaffe while at the same time launching their own bombing raids on Germany. Hitler also opted for a 'starve-them-out' policy, attempting to use submarine warfare to disrupt the British overseas lifelines in the Battle of the Atlantic, which was eventually turned by the Enigma discovery of German codes by Station X at Bletchley Park. The Germans now had submarine bases in Norway and France. Invasion was anticipated in Britain, but Hitler and his advisers drew back from attempting to cross the English Channel until the British air force could be neutralised first. As a result, the Battle of Britain was fought in the air, not on the beaches. Instead, the Germans launched daylight raids on ports and airfields and later on inland cities.

Although priority had turned to the defence of Britain, Churchill was well aware that the spirit of attack had to be maintained, if for no other reason than to boost the morale of the British nation at its darkest hour. This, he felt, could be achieved only by raids on German-held coastlines. In what his biographer Martin Gilbert described as 'searching for some means of a British initiative' and not least a measure of success that could be relayed to the British people clinging hopefully to his every word, Churchill demanded action of the military Chiefs of Staff, as a minute logged on 3 June 1940 makes clear:

> The completely defensive habit of mind which has ruined the French must not be allowed to ruin all our initiative. It is of the highest consequence to keep the largest numbers of German forces

> all along the coasts of the countries that have been conquered, and we should immediately set to work to organise raiding shores on these coasts where the populations are friendly. Such forces might be composed by self-contained, thoroughly equipped units of say 1,000 up to not less than 10,000 when combined.
>
> Enterprises must be prepared with specially trained troops of the hunter class, who can develop a reign of terror first of all on the 'butcher and bolt' policy but later on, or perhaps as soon as we are organised, we should surprise Calais or Boulogne, kill and capture the Hun garrison and hold the place until all preparations to reduce it by siege or heavy storm have been made, and then away.
>
> The passive resistance war which we have acquitted ourselves so well in must come to an end. I look to the Chiefs of Staff to propose me measures for a vigorous, enterprising and ceaseless offensive against the whole German occupied coastline. Tanks and AFVs [Armoured Fighting Vehicles] must be carried in flat-bottomed boats out of which they can crawl ashore, do a deep raid inland, cutting vital communication and then back, leaving a trail of German corpses behind them.

Expanding later on these thoughts to his Chief of Staff General Hastings Ismay on 5 June, he suggested the creation of detachments of troops to be used in forays against enemy-held coastlines 'equipped with grenades, trench mortars, Tommy-guns, armoured vehicles, and the like, capable of landing on friendly coasts now held by the enemy . . . it is essential to get out of our minds the idea that the Channel ports and all the country in between them are enemy territory'.

The Joint Chiefs of Staff were presented with these ideas at a meeting the next day, 6 June, and the reaction was swift. Three days later the War Office dispatched an urgent call to all commands requesting the names of 40 officers and 1,000 other ranks to join a 'mobile force', and it was made known that the ultimate numbers required might reach 5,000. The force was now formally to be recognised as Commandos, and already in the pipeline was a plan to set up ten commando units as soon as possible. MO9 Branch was created to take responsibility for planning all Commando operations, and on 12 June Sir Alan Bourne, Adjutant-General of the Royal Marines, was asked to become Commander, Offensive Operations – forerunner of Combined Operations – with a mandate to call on the cooperation of both the Royal Navy and the RAF.

On 13 June the detail and specific ethos of Commando formations and operations were outlined in a top-secret memorandum from Major-General R. H. Dewing, Director of Military Operations and Planning, and those familiar with the Boer Commandos would not be long in recognising the similarities:

> It is intended to form the irregular volunteers into a number of commandos, and I have prepared this memorandum to explain the purpose of the new force and the way in which it is proposed to organise and employ it. The object is to collect together a number of individuals trained to fight independently and not as a formed military unit. For this reason a commando will have no unit equipment and need not necessarily have a fixed establishment. Irregular operations will be initiated by the War Office. Each one must necessarily require different arms, equipment and methods, and the purpose of the commandos will be to produce whatever number of irregulars required to carry out the operations. An officer will be appointed by the War Office to command each separate operation, and troops detailed to carry it out will be armed and equipped for that operation only from central sources controlled by the War Office. The procedure proposed for raising and maintaining commandos is as follows:
>
> One or two officers in each Command will be selected as Commando Leaders. They will each be instructed to select from their own Commands a number of Troop Leaders to serve under them. The Troop Leaders will in turn select the officers and men to form their own Troop. While no strengths have yet been decided upon, I have in mind commandos of a strength of something like 10 troops of roughly 50 men each. Each troop will have a commander and one or possibly two other officers. Once the men have been selected, the commando leader will be given an area (usually a seaside town) where his commando will live and train while not engaged on operations.
>
> The officers and men will receive no Government quarters or rations but will be given a consolidated money allowance to cover their cost of living. They will live in lodgings etc. of their own selection in the area allotted to them and parade for training as ordered by their Leaders. They will usually be allowed to make use of a barracks, camp or other suitable place as a training ground. They will also have an opportunity of practising with boats on beaches nearby.
>
> When a commando is detailed by the War Office for some specific operation, arms and equipment will be issued on the scale required, and the commando will be moved (usually by separate Troops) to the jumping-off place for the operation. As a rule the

> operation will not take more than a few days, after which the commando would be returned to its original 'Home Town' where it will train and wait, probably for several weeks, before taking part in another operation. It will be seen from the above that there should be practically no administrative requirements on the Q [rations and domestic supplies] side in the formation or operation of these commandos. The A [personal matters] side must of course be looked after, and for this purpose I am proposing to appoint an administrative officer to each commando who will relieve the commando leader of paperwork. This administrative officer will have permanent headquarters in the 'Home Town' of his commando. The commando organisation is really intended to provide no more than a pool of specialised soldiers from which irregular units of any size and type can be very quickly created to undertake any particular task.

In the coming weeks, it was proposed to form 12 Commando units of 500 men apiece, utilising to some extent some of the independent companies that remained active for the time being. Churchill, meanwhile, was impatient for action, and having mentioned in his comments to the military leaders that he foresaw some attacks on the French coastline in the very near future the planners prepared to do exactly that – before the month of June was out. It was a mistake, as will become apparent as the account of the first two operations supposedly in the Commando mould now unfolds.

The true Commando units were not even at the recruitment stage when the operations were being planned, so a newly created No. 11 Independent Company was raised on 14 June, the day after Major-General Dewing issued his directive on the structure of Commando units. It was clearly intended as a trial run, using as many elements as possible from the Commando ethos. The company was made up in part by new Commando volunteers and the old No. 9 Independent Company (IC), which – as previously described – had been hanging around in Scotland after sailing up and down the Clyde when their mission to Norway was aborted. The company was to be commanded by Major Ronnie Tod and would consist of 25 officers and 350 other ranks who would now have the dubious honour of carrying out the first of the officially designated Commando operations. 'Dubious' was the word, because once again everything was done in a hurried fashion, and rather than having weeks of training – as later became the norm – the men were once again thrown straight into the ring to start punching. Not surprisingly, things did not quite work out as planned. Ronald Swayne, who had switched from the now-disbanded No. 9 IC from Ross-on-Wye to the new No. 11 Company, remembered:

> We were still parked up in Glasgow, waiting for something to do after the evacuation from Norway, when we suddenly got an instruction that some officers who could select their own men were to report at a certain place in Southampton. They were to find their own way there. They were to billet themselves. They were to be paid thirteen shillings and fourpence [69p] a day to an officer and six and eightpence [36p] a day to a soldier. And they were to take the most extraordinary combination of equipment, including a sniper's rifle, and this we did. I and two or three of my friends went off all travelling quite independently to Southampton. I suppose that was the real beginning of the Commandos. It was a very strange formation and was intended, we learned in due course, to make raids immediately on the French coast. I and my friends put up at the Royal Court Hotel in Southampton, very comfortably. We trained very hard for a few days and eventually we were told to go to a hotel in Whitehaven, which we did by train. We were there briefed by Jumbo Lester, who was then a colonel, a marine with a huge moustache. He gave us a short talk on C-boats, which we were apparently going to use to get us across the Channel. The C-boat, or crash-boat, was a very small motor-driven craft which had a little cabin and was built to RAF specification to pick up pilots when they ditched into the drink. It only did about eight knots, and it didn't have room for more than about twelve of us.

On the afternoon of 23 June, Major Tod received his sealed orders from the War Office for what was to be known as Operation Collar, instructing him to split his force and make landings the following night at four different points on the French coast. The object of the trip was to reconnoitre the defences, capture a few Germans for interrogation and do anything else that might be of use. It was also to provide much-needed morale-boosting headlines that little less than three weeks after the evacuation from Dunkirk, British troops had landed on French soil to harry the German invaders.

In the haste to achieve these objectives, training, planning and intelligence went awry, and the raid fell into disarray almost from the outset. The first of Tod's setbacks occurred when he learned that only nine of the promised twenty C-boats were available. He decided to split his force into five; they would land in separate locations between Boulogne and Berck. The landings would take place at midnight, and the parties were to be allowed 80 minutes ashore. Ronald Swayne, leading one of the groups, continued his account:

> My men, armed to the teeth, were going to throw hand grenades and generally beat up a hotel at Merlimont Plage in Le Touquet [thought to have been requisitioned by the German army]. Things didn't go well on the way out, and we arrived rather late; it was almost beginning to get light. We went ashore and headed to our objective and found the hotel was all boarded up. There was nobody there. We wandered round a bit, looking for Germans, and then gave it up because time was running out. We went back and discovered there was no boat. It was hanging around some distance offshore and we couldn't make contact with it. Some Germans turned up whom we killed and that created a bit of a noise. I'm afraid we bayoneted them. I was armed with a .38 revolver. I'm sorry to say that I forgot to load it on this occasion. So I hit one of the Germans on the head with the butt of my revolver. My batman bayoneted one, and I grappled with another and we killed them. It wasn't really very serious soldiering, I'm sorry to say. And, of course, because we were being rushed – it was unpardonable, really – we never got their identity papers, which was very inefficient of us. We also lost a lot of our weapons.
>
> Then some more Germans appeared in the sand-dunes, and I needed to get the men away fast, so there was nothing else for it but to swim out to the boat, still parked some distance offshore. I swam till I was very nearly exhausted. I was wearing breeches, which are not best suited for swimming long-distance. I demoted the naval officer on the boat and put the coxswain in charge; he took the boat in closer and eventually we got everybody off. It was a rather amateurish affair, although I'm glad to have mentioned it because I later read an account by a Cardiff academic who wrote about us throwing hand grenades into a café filled with French girls and German soldiers, which was quite untrue. It was the last thing we would have done. We might have waited outside and beaten up the Germans, but I don't think we would have, I hope, bombed civilians in a café. At least we wouldn't have done, not at that stage in the war. I think he must have picked it up verbally from somebody who wasn't there.
>
> The truth was, we didn't see many Germans. They only came in some force once we'd made a noise. We were under quite heavy fire when we withdrew and there was a lot of stuff flying around. We all got away quite safely.

Swayne's group probably saw the most action. Other members of Tod's force landed at Boulogne, Hardelot and Berck, while the fifth party arrived so late the men were unable to land at all and returned to base without exiting their dangerously waterlogged boat. Those who managed to get ashore carried out their reconnaissance of the scene, noting such inconsequential features as searchlights and a sealed-off area that they thought might be fuel or ammunition dumps. There were a couple of skirmishes with Germans, and the British parties came under machine-gun fire as they departed. The 'raid' had its moments, as recalled by Ernest Chappell, who was with the CO's party:

> Major Tod had a Thompson sub-machine-gun, one of these things with a drum magazine attached. He was scouting along a road when he heard some cyclists coming up behind; they were Germans on bikes. As he swung around to confront them, machine-gun at the ready, the magazine fell off the gun. Fortunately, the German cyclists were equally as surprised as him. They turned round and went the other way. There were no shots fired there. So we came back home. The raid didn't achieve much, other than being looked upon as experience and training. But it was quite a boost to morale, both to ourselves and, I suppose, to the civilian populace to read headlines in the newspapers that our forces had landed on the Continent less than three weeks after the evacuation of the BEF. Never mind what we'd done or hadn't done. We'd landed and we'd come back again without any casualties. The possibility was there of us doing it again, and that, in my view, was what it was all about.

A second raid launched from the South Coast of England was on the cards within the week, again inspired by Churchill himself. On 2 July, two days after the Germans had landed in the Channel Islands, he dictated a note to General Ismay: 'If it be true that a few hundred German troops have landed on Jersey or Guernsey by troop carriers, plans should be studied to land secretly by night on the Islands and kill or capture the invaders. This is exactly one of the exploits for which the Commandos would be suited.'

Again, there was a swift response to the Prime Minister's suggestion. Later that day the War Cabinet approved proposals for a raid by 140 men – who would include some of the men who had already crossed the Channel – to land in the Channel Islands. Their object was to destroy German aircraft and facilities on the airfield there, capture some German soldiers for interrogation and kill as many as possible. This was to be the first true test under the Commandos' banner. The first unit of the dozen units in the process of formation had completed its recruitment, although there had been little time for its commanding officer, Major John Durnford-Slater, to put his men through their paces. No. 11 Independent Company, formed for the Boulogne raid, was to join the party, which was codenamed Operation Ambassador. Durnford-Slater had just ten days to train his men, with the attack planned for arrival in the Channel Islands at midnight on 14 July.

Three separate points on the southern coast of Guernsey were selected; two parties from No. 11 Independent Company were to go ashore at Le Jaonnet Bay and Pointe de la Moye, while forty men from the newly formed No. 3 Commando under Durnford-Slater were to land at Moulin Huet Bay. The raiders were to be carried towards their target aboard two destroyers that would take them as close as they could, and thereafter the men would clamber into small boats to go ashore. The convoy set off from Dartmouth at 1845 hours, and on board once again was Ronald Swayne with No. 11:

> In theory, it was beautifully planned from an army point of view, but the naval preparation was very inefficient, partly due to inadequate equipment, of course.[1] Two parties from No. 11 were to be landed, one as a diversion on a fairly easy beach on one side of the island. They were to land first. And then my party was to land at the cliffs, which were quite close to the airport. We were to bomb the hotel, which was full of German pilots, and blow up the aircraft on the airport. There were quite a lot of planes on the airport. We had Channel Islanders with us who could guide us there from a place on the cliffs, where we were to land, up a path to the top of the cliffs. There were also Nissen huts occupied by German soldiers round the airport, and we were to attack them and blow up the planes. It was a wonderful idea, and it could have been a very, very clever raid. But the means of transport were absolutely hopeless. They towed our little boats – gigs, whalers – behind the navy destroyers which transported us. We were to use these boats to row ourselves ashore. Several crash-boats also went under their own steam. Well, the little boats were damaged en route. When we got out from the destroyers, they were all leaking; some were useless. One by one we were

[1] An official report later noted that RNVR volunteers manning the boats had forgotten to adjust their compasses, which were 'many degrees out', so landing places had to be 'guessed at'.

> transferred best we could, and eventually I was in a crash-boat which was full of water. All this had taken so long we'd arrived at the rendezvous late, and it was nearly daylight.
>
> Fortunately, there was a bit of sea mist but in the end it was quite impossible to attack; the situation for my party was quite hopeless. Furthermore, we'd lost sight of the destroyer which was to take us home. We decided to put into Sark and get rid of half the people on board so that there was a chance of this boat getting back to England under its own steam, under its own power. We were making for Sark when the destroyer spotted us. Somebody on the destroyer – it was a soldier, actually – spotted something in the water. He knew that there were people missing and he drew the attention of a ship's officer and he got the ship diverted to pick us up and we got home.

One of the raiders' boats did indeed get to Sark – by mistake. Faulty navigation caused the boat to miss Guernsey altogether and arrive at an impossible landing site: high cliffs with no beach. As Ernest Chappell recalled succinctly: 'We just couldn't believe it – we were only on the wrong bloody island. That's how silly it was. So our mission was over in half an hour. We just rowed the boat back to find the destroyer.' Of the three groups in their motley collection of small boats, only Durnford-Slater's party reached the correct beach on time, but in fact they were unable to get their boats close enough for a dry landing because of the rocky terrain. They had to wade through shoulder-high seas and clamber over a jagged rock face, which delayed the start of their allotted time ashore. They went as far as they could within the time-frame but found no sign of any Germans, and the commanding officer had to order a return to the boats for a rendezvous with the destroyer that was to take them back to England.

As they did so, they were spotted by a German machine-gun post that opened fire. By then, too, the seas were very heavy, and with waves crashing into their landing area it was impossible to get their boats in closer than 75 metres, so the men had to swim out in rough seas with their equipment tied around them. One was drowned almost immediately; three others who had previously insisted they were able to swim, couldn't. They tried to wade to the boats but they could not make it and gave up. They were given extra French francs from the emergency survival money, and Durnford-Slater promised to have them rescued from a given rendezvous two nights later. In fact, the three were captured later that night and even if they had stayed free they would not have been rescued because Naval Command refused to sanction a rescue attempt.

Nothing was achieved by this raid, other than experience, with only one party actually getting ashore. Ronald Swayne summed up:

> We were very lucky to get home. Again, it was an amateurish business but a necessary part of the learning. We weren't remotely ready. We didn't know what kind of boats we wanted, we didn't know how to do it, there wasn't a proper planning body to prepare for these raids and there was still a fearful shortage of weapons. We were armed with .38 revolvers, which I always thought a very poor weapon. We had a few Tommy-guns later, but there was a shortage of ammunition for teaching the soldiers to shoot. Some of them had hardly done any practice on the range.

According to Lord Lovat, soon to become a leading figure in the Commando movement, 'the Prime Minister was not amused by this tomfoolery and laid it on the line in no uncertain terms. The sailors got a reprimand and Churchill ordered an immediate reorganisation.'[2] On 23 July in a note to Anthony Eden, he wrote that it would be 'most unwise to disturb the coasts of Europe by the kind of silly fiascos which were perpetrated at Boulogne and Guernsey. The idea of working these coasts up against us by pinprick raids and fulsome communiqués [published in the newspapers and broadcast by the BBC] is one strictly to be avoided.' But he went on to say that there would be no objection to 'stirring up the French coast with minor forays' during the winter of 1940–41 but that much larger incursions had to be contemplated, properly planned and executed, for the spring and summer of 1941 and beyond. It was a good enough signal to complete the formation of the 12 Commando units as soon as possible and, as a matter of priority, to organise some realistic training in the most difficult of surroundings to equip the soldiers with strength of will for what lay ahead.

While most of the independent companies were in time disbanded, they still provided a ready pool of manpower, and every man who opted to join one of the new Commando units was re-interviewed by the recruitment teams. Fresh calls went out for volunteers, and each and every one who applied would get a personal interview, guaranteed. A predetermined selection procedure would be ruthlessly applied, first weeding out those considered unsuitable at the interview stage and then putting them through a tough initial training programme that many would fail and be RTUd.[3] Speed, once again, was of the essence,

[2] Lord Lovat, *March Past,* Weidenfeld & Nicolson, London, 1978.
[3] Returned To Unit from which originally recruited.

demonstrated by the prompt manner in which the 6,000 other ranks were recruited: all bar one of the 12 Commandos had been filled up by the beginning of August, each with 500 men. To give further impetus, Churchill appointed his old friend Admiral Sir Roger Keyes (later Baron) as head of the newly named Combined Operations, which would replace the oddly titled Offensive Operations, which had at its helm Sir Alan Bourne. Bourne had also remained Adjutant-General of the Royal Marines and was clearly under great pressure. He agreed to relinquish his Marines job and become deputy to Keyes. They installed themselves in a building in Richmond Terrace, Whitehall, which remained the headquarters for Combined Operations for the rest of the war.

Keyes himself was the architect and hero of one of the British military's few Commando-style operations in the First World War. In 1918, when commanding the Dover Patrol, he led raids on German naval bases at Zeebrugge and Ostend, effectively blocking the ports for use by U-boats, which historians later regarded as classic Commando operations. He later became Commander-in-Chief in the Mediterranean and on retiring from the Royal Navy became Conservative MP for Portsmouth.[4] But, being an excitable sort of chap, he had many detractors in Whitehall. Even Churchill himself had rebuffed Keyes' own suggestion in April that he should personally lead a special force to capture Trondheim during the Norwegian crisis with a letter which began, 'Dear Roger . . . It astonishes me that you should think that all this has not been examined by people who know exactly what resources are available and what the dangers would be . . . it is not open to me to make appointments on the grounds of friendship.'

In meetings prior to his arrival as head of Combined Ops, Keyes was said by Lord Dunglass[5] to be 'so excited . . . as to be almost incoherent, and apparently heading for a brainstorm, so he may generally have been misunderstood'. Roger Keyes began what would ultimately become a controversial reign over Combined Operations in an arena of some mistrust and little cooperation from the conventional military top brass who by then were deeply concerned with the Battle of Britain. The army, navy and air force were all deployed on the protection of British shores, ships and cities.

[4] He was a personal friend of Churchill who wrote the foreword to Roger Keyes' book, *Adventures Ashore and Afloat*, published just before the outbreak of the war, noting 'when barely 27 he commands the torpedo boats which cut out Chinese destroyers near the Taku forts'.

[5] Chamberlain's private parliamentary secretary in negotiations with Hitler; later, as Sir Alec Douglas-Home after relinquishing his peerage, to become British Prime Minister in 1963.

Even so, as far as the Commandos were concerned, Keyes abided by Churchill's wishes and sent rattling memos flying from Richmond Terrace to get the whole concept on to a permanent footing with real training under conditions that tested the endurance of the men against all the possible hazards they were likely to meet. One training school for officers had already been established, using the Scottish estate of Lord Lovat in the west Highlands. It was formed almost as a private enterprise by a group of officers who had served in the Scots Guards together in Norway. It was originally Major Bill Stirling's idea. He and his younger brother David were cousins of Lovat and were very familiar with the countryside, having been visitors there for many years. They obtained authorisation from the War Office to set up a new training school on the Lovat property, which included six deer forests with lodges and covering a landmass for training purposes running to some 80,000 hectares. Lovat himself was sent on ahead to requisition all properties astride the Fort William to Mallaig road and the railway line that would be useful. Once installed with Bill Stirling as chief instructor, they began with cadre courses for junior officers, who began arriving in the first week of June 1940. Their base was Inverailort Castle, 25 miles west of Fort William. It was a large, square building, grey and gloomy, rising like an apparition from the rolling mists in the dark dawns. Built on the shores of Loch Ailort, it provided all the opportunities for Commando training, including amphibious operations supervised by a naval commander. Lord Lovat himself provided men from his estate, stalkers who were great rifle shots and expert telescopic men, who turned up in their civilian plus-fours to give instruction.

Numerous officers came through the school, including several who were to become famous in the Commando operations, like David Stirling himself, then a 24-year-old second lieutenant, and 'Mad Mike' Calvert, then 25, who became one of the great pioneers of guerrilla warfare. His personal exploits included assisting Major-General Orde Wingate in the famous Chindits in the Burma campaign, along with his association with Stirling's SAS. It was Calvert's words, spoken often enough, that were ingrained into the psyche of many a volunteer to the Special Forces later: 'The main job of the soldier is to kill people. As a guerrilla you don't achieve anything by just being present. No regular force of any nation in the world is really frightened of guerrillas unless they can see the results in blown bridges, their friends being killed or trucks being ambushed.'

The school was a great success. Ronald Swayne, by then with No. 1 Commando, remembered:

> The whole Commando world attracted some very strange characters to it. Some were extremely confident, useful people who were very successful later in the war. Several were just total eccentrics, and the training centre became a sort of home for some of them. We had somebody up there called Commander Murray Levick, who tried to make us live on pemmican. We all pretended we lived on it but we used to go off and get a good meal of venison or something down in the village after supper. Some of them added a lot of imagination to training and I think helped and developed the eventual character of the Commandos.

Talk among returning officers was of great tests of physical endurance, a trial of strength in every respect. Roger Keyes had heard of it by the time he became Director of Combined Operations and set up a Combined Training Centre for amphibious warfare at Inveraray. Winston Churchill himself, meanwhile, kept up the pressure. Memo to Anthony Eden, 25 August:

> If we are to have any campaign in 1941 it must be amphibious in its character and there certainly will be many opportunities for minor operations all of which will depend on surprise landings of lightly equipped mobile forces accustomed to work like packs of hounds instead of being moved about in the ponderous manner which is appropriate. These have become so elaborate, so complicated in their equipment, so vast in their transport that it is very difficult to use them in any operation in which time is vital. For every reason therefore we must develop the storm troop or Commando idea. I have asked for 5,000 parachutists and we must also have at least 10,000 of these small 'bands of brothers' who will be capable of lightning action. In this way alone will those positions be secured which afterwards will give the opportunity for highly trained regular troops to operate on a larger scale.

If any further recommendation was needed, it came from his own son Randolph, who immediately volunteered to become a Commando. James Sherwood, from Southport, Lancashire, who had lately been a dispatch rider with the RAMC, scuttling around Kent on a motorbike delivering messages, volunteered for 'special duties' at the same time and ended up in 8 Troop of No. 8 Commando, raised at Windsor by Captain Godfrey Nicholson. Sherwood recalled:

> We were among the first batch in October 1940. We were thrown in at the deep end and at that time each Commando set about training

> according to its own ideas. It started out on the basic premise of weeding out those who couldn't stand the pace. Within 24 hours of arriving there, we formed up complete with our gear, full pack: rifle, respirator, tin hat and the rest of the paraphernalia. We were headed by Randolph Churchill, who looked to us to be very fat and unfit but who proved himself as capable of taking on anything that we had to do. What we could do, he could do. I remember the perspiration poured off him. He must have lost about a stone in no time at all, on what turned out to be high-speed marching which just stopped short of running. I think the aim was to do something like seven miles an hour which, with all the gear, was some going, especially for people from units like my own who had been sitting in vehicles for most of the war.
>
> We belted out along the road northwards by the loch with Randolph at the head going hell for leather for about an hour. Then, after a brief stop, we turned around and did the same thing coming back. Those who couldn't take it just fell out by the roadside and were returned to their unit the next day and never seen again. The rest of us were just doubled up with pain and Randolph was still there, still pouring buckets of sweat, and then he shouted, 'Pick up the step' and some rebellious spirits shouted back, 'Bugger off!' That was our first day.

The training of No. 8 Commando was similar to that eventually established for all units, although at that stage the regime was still fairly basic: forced marches and heavy pressure. The Commandos were all given preliminary training at different concentrations along the Firth of Clyde, while much-needed training locations for junior naval officers in the art of manoeuvring assault landing craft were set up at Warwash on the Hamble River and elsewhere around the Portsmouth estuary. Incredibly, however, a formal training centre for Commandos did not come into being until February 1942.

In the meantime, training remained on an ad hoc basis, with each unit responsible for its own training regime, and to complete the story of 8 Troop 8 Commando, the whole unit was put aboard the recently acquired Commando assault ship HMS *Glengyle*[6] and sailed into the Clyde to Gourock, where they offloaded and were bussed en masse to Largs, on the remote Ayrshire coast. The townspeople had never seen soldiers before and were now expected to provide billets for the

[6] One of three peacetime *Glen* ships assigned to the Commandos in 1941; they were fast cargo vessels of around 10,000 tonnes. The others were *Glenroy* and *Glenearn*.

Commandos in accordance with the system. They were accommodated in threes and fours. As James Sherwood recalled: 'The families turned somersaults for us. They couldn't do enough. The families on whom we were billeted looked after us as if we were their own.'

They remained in the seaside town for about a month, training and marching, and then the whole Commando was moved to the Island of Arran, again in civilian billets, in the village of Lamlash. But bad news was on the way. Sherwood again: 'Within a day or two of arriving there, Godfrey Nicholson, who was in charge of 8 Troop, called us together and told us, to our fury, that our standard of training wasn't considered sufficiently high by the command of 8 Commando and we were going to be disbanded – which meant only one thing: RTU. The men were very angry, because they considered the training to be the responsibility of the officers. There was a lot of offensive language used on parade, which in other circumstances would have been acted against.'

It so happened that at Lamlash at that time was one Lieutenant Roger 'Jumbo' Courtney, an extrovert character of some repute, who had just been given permission by Lieutenant-Colonel Bob Laycock, commanding officer of No. 8 Commando, to form an experimental Folboat Section. Courtney was a wild, adventurous man of 40 who between the wars had been a professional big-game hunter and gold prospector and who, when he married at the age of 36, took his wife Dorrise aboard his two-seat collapsible canoe named *Buttercup* and paddled down the Danube. In another adventure, he paddled the Nile from Lake Victoria to the delta with only a sack of potatoes and an elephant spear on board. He convinced Roger Keyes of the versatility of clandestine attacks by canoe-borne Commandos by staging a mock raid on the *Glengyle*. Still dripping wet, he burst into a high-level Commando conference onshore and dropped a gun cover on the table to prove he had got aboard the ship unnoticed. The upshot was the foundation of the Folboat Section, which in turn became the launchpad for the Special Boat Service, which remains in existence today under the Royal Marines' banner.[7]

James Sherwood and half a dozen other disgruntled members of the disbanded 8 Troop marched over to see Courtney at his HQ in Lamlash. Most were taken on to join other recruits, numbering 12 in all, who were to form the basis of an important and swashbuckling crowd. The Folboat Section recruits made their name being launched from submarines with canoes, paddling ashore and landing on enemy-held shores, blowing up

[7] As told in the complete history of the unit in John Parker's critically acclaimed *SBS: The Inside Story of the Special Boat Service*, published by Headline, 1997.

vital installations and committing other acts of sabotage from which several did not return. Several others from 8 Troop, including Randolph Churchill, remained with 8 Commando as it prepared to join the fray elsewhere, destined eventually for the Middle East.

Chapter Three

Better Results

By early October another reorganisation of the Commandos to meet the aspirations of Winston Churchill was under way. The remnants of the independent companies, many of whom, to use Lord Lovat's description, still 'squatted dejectedly about the glens in muddy tents', joined together with Nos. 5, 6 and 11 Commandos to create three Special Service Battalions. Each battalion would have two 500-man companies, each formed from an existing Commando. Two more battalions would then be raised from Nos. 3, 4, 7 and 8 Commandos. They would make up a 'Special Brigade' and reorganisation was completed by early November.

The changes meant the final disappearance of the independent companies. Their disbandment and the manner in which many men were sent back to their original regiments caused bitterness, as reflected in a diary note from Major Bill Copland, as he gave up No. 4 Independent Company: 'And so, into another new organisation goes the cream of the Independent Companies, which were born in muddle, suffered in the chaos of Norway, wasted valuable time at home when tasks which other troops botched could have been done well by them . . . I hope that the new staff will be more worthy of them than ever was the staff which controlled their Norwegian destinies.'

Ronald Swayne recognised the difficulty:

> We got rid of the officers and men whom we didn't think were up to scratch, either not fit enough or a bit wet or something or other – or we just didn't like 'em. The Commandos tended to be clubs. When we had a vacancy, you could put up suggestions to the colonel. Opinions would be sought, like the blackball system, and each Commando had its own character. No. 8 Commando – Laycock's outfit – had apparently been recruited from a bay window in White's

> Hotel. The officers and men and NCOs of all Commandos were fairly hand-picked. We were, most of us, very fit physically, played games and that sort of thing but enjoyed long hauls over the mountains. We did masses of it. It was nothing to do 20 or 30 miles in a day, cross-country. We did become immensely fit.

There was one other interesting development which again was the result of Winston Churchill's demands for versatile go-anywhere units. As well as Commandos, he had asked for the formation of a parachute regiment with a strength reaching at least 5,000. The RAF were initially to oversee its creation but were already fully engaged and overstretched to the absolute limit during the Battle of Britain. There were not enough planes to carry bombs, let alone for people to jump out of. The project was thus handed to the army, with joint participation of the RAF training instruction, and consequently No. 2 Commando was transferred en masse to form the nucleus of the first ever parachute troops in Britain under the title of 11 Special Air Service Battalion. They moved to Ringway, near Manchester, where the nation's first parachute training centre was being built. It would eventually become the 1st Parachute Battalion and later the Parachute Regiment, leaving the Special Air Service title to be snapped up by another young swashbuckling officer in the Commandos, one David Stirling – who was soon to have great ambitions which, said many, were above his station but resulted in the creation of the SAS. Within months, therefore, out of the Commando idea – and Churchill's personal interest – the foundation stones were laid for four fledgling but crucial elements that were dramatically to affect the future outcome of the war, and impact on military thinking for the rest of the century: the Commandos, the Parachute Regiment, the Special Boat Service (SBS) and the Special Air Service (SAS).

As the new year of 1941 dawned, it was still a hit and miss affair. In January Lieutenant-Colonel Bob Laycock received orders to put together hurriedly what was officially known as Force Z. The movement was classified top secret; no word of its destination was to leak out. It was mounted by the Chiefs of Staff with the specific intention of capturing the island of Rhodes, to provide the British with a staging-post in the eastern Mediterranean. The force of 100 officers and 1,500 other ranks consisted of Nos. 7, 8 and 11 Commandos, A Troop from 3 Commando and Roger Courtney's Special Boat Section, as it was by then known. They sailed from the Island of Arran on 31 January in the Commando ships *Glenroy* and *Glengyle* and en route Laycock was informed by the War Office that with immediate effect his group was to be renamed Layforce and that any mention of the word Commando was expressly forbidden. They were also to be joined by the third *Glen* ship, *Glenearn*, carrying specialist teams from the Mobile Naval Base Defence Organisation, formed under the auspices of the Royal Marines a few months earlier for the security of naval bases.

They were due to arrive at Suez in early March and link up with two other Commando units, 50 and 52 Middle East Commandos, formed *in situ* in June 1940. Those who remained in Britain were indignant and moaning about the lack of action, but in the event it was just as well. Some difficult times were already signposted for the Middle East, and we will catch up with the activities of Layforce in due course. Those who stayed back had been earmarked for numerous raids in the early months of the year, most of which were either cancelled or aborted at the point of readiness. They included training work-ups for Operation Workshop, to land a Commando force on the island of Pantelleria, off the coast of Sicily, and Operation Brisk, to seize the Azores. Neither went ahead. In February, however, the green light stayed on and the Commandos were planning their first raid since the Guernsey fiasco, one that would take them back to the Narvik area of Norway. It was codenamed Operation Claymore, and the target was the Lofoten Islands, which lay 850 miles north of the Orkneys, near Narvik. The islands were important to the Germans because there were oil installations to serve shipping but more important were the factories that processed herring and cod extracts, from which fish oil was produced in large quantities for a variety of uses, including the manufacture of nitroglycerin for high explosives. There were also smaller manufacturing plants that produced the oil in bottles for vitamins to help German troops withstand the cold of the Arctic Circle.

The orders were straightforward: to blow up the oil installation and any enemy shipping they could lay their hands on, capture prisoners and Norwegian quislings, and carry back any Norwegians who wished to volunteer for the British-based Norwegian forces. Nor was there to be any mercy shown to shipping, which carried the local products under duress from the Germans. 'They were to be burned without compunction,' ordered Lord Lovat, who had just been posted to No. 4 Commando, which would combine with 3 Commando, each with 500 men, plus a force of 50 Norwegian troops for the raid. After a couple of false starts following disagreements in London over tactics, the raiders sailed from the Clyde up to Scapa Flow in the Orkneys for an exercise and then on to the Faeroe Islands for final briefing and training. The naval vessels HMS *Queen Emma* and *Princess Beatrix*, darkened and packed to the gills with troops and equipment and top heavy with land craft, slung, bounced and buffeted across the North Sea. Many of the soldiers –

especially the first-timers – were mightily sick. Two destroyers provided the escort through waters often alive with German U-boats, and the submarine *Sunfish* had gone on ahead to provide navigational aid. A naval commander who came aboard the *Emma* at Scapa Flow to act as navigator knew the waters well – and the dangers. No one was aware until later that he was one of the survivors of the aircraft carrier *Royal Oak*, torpedoed the previous autumn in the Flow with the loss of 800 lives.

The convoy reached the Lofoten Islands in the early hours of 4 March 1941, and the force, armed to the teeth and faces blackened, was ferried ashore under freezing conditions in a dozen landing craft, each carrying 35 men. They landed on a mission that was to be the largest and most successful since their foundation, albeit without a great deal of opposition. The assault landing craft sped in to land the scouts and patrols who went on ahead to pinpoint the factories and installations for the demolition men to deal with. The fighting patrols, with plenty of hefty weapons, took up their positions in case of trouble. They commandeered what vehicles they could; one unit from 4 Commando, for example, had an old fish lorry with a Bren gun mounted on the cab to lead their ramshackle convoy. Lord Lovat recalled that they soon ran into trouble, although with nothing they couldn't handle:

> We turned a blind corner and ran slap into a column of Germans. A Norwegian officer pulled the trigger of his revolver in shocked surprise. The heavy bullet smashed into the dashboard where my kneecap had been resting, filling the cabin with cordite fumes. The lorry swerved and hit the parapet of a low bridge . . . which threw the Bren gunner and his number two on to the hard-packed snow. The marchers turned out to be unarmed sailors of the German merchant marine . . . one of them, more courageous than his fellows, dropped his satchel over the parapet into the creek below.

In the scuffle that followed a German who turned out to be the skipper of the ship from which they came was knocked over the parapet into the creek. Lovat said that the Bren gunner, angered because his dentures were broken in the crash, was all for taking a pot-shot at the swimmer and was urged on by the German skipper's colleagues who hated him because he was an ardent Nazi. Eventually, the man was hauled from the water and his log in the satchel retrieved. The patrols moved on and into town. By then, the demolition men were already at work, within an hour of landing. Bernard Davies, a young veteran of the Dunkirk evacuation, was on his first raid with 4 Commando, which he joined within three weeks of his return from France:

> I was stationed at Chichester with the 10th Infantry Brigade, back from Dunkirk about a month. We were a bit fed up – no weapons to speak of, continually doing routine training and so on – and then one day a smart young officer arrived and asked for volunteers for a special force. About 20 of us volunteered. He formed B Troop of 4 Commando and on 5 July we all assembled down at Weymouth, where we began some very serious training for three or four months. We had no idea what it entailed other than that we were going to be engaged in raiding the enemy on the coast regions, which appealed to us foolish boys. We wanted action; get our own back for the humiliation. The troop I was in was completely made up of regular soldiers, fine fellows in every way: morale, temperament, everything. They were very selective. If you didn't fit you were kicked out fairly quickly. They came from every regiment you could think of; they wanted a variety of skills.
>
> The excitement of the Lofoten raid was just what we volunteered for, although I have to say we were met with very little serious opposition. There were mostly Luftwaffe troops there, technicians, and we captured quite a number of them. We destroyed a lot of shipping in the harbour, took very few casualties, on our own side at least. We slaughtered a few Germans, quite unnecessarily in my view; they would have surrendered anyhow. Some were trying to get away. We did have some bloodthirsty characters with us. Fortunately, they were few and far between. It was an image some people have of Commandos, but it's the exception rather than the rule. Perhaps, one in ten were bloodthirsty; the rest of us were quite normal. I could have killed quite a few Germans; I personally saved a lot of Germans from being killed. Some of our guys would kill the wounded if necessary.

Norwegian locals emerged from their homes as soon as they realised that the British had arrived. They watched the first of the explosions as the factories were blown to smithereens and then, as the soldiers moved into the administration areas, quislings who cooperated with the Germans were pointed out and arrested by the British troops. According to Lord Lovat, there was a thunder of applause from a crowd of patriots who were waiting at the harbour – 315 of them, mostly young men and women in ski-pants and windcheaters with a few belongings slung into rucksacks, managed to hitch a ride to the UK, and freedom. They tagged along with the raiders as they clambered aboard the landing craft to be ferried back to the waiting ships in open water, with Lofoten providing a backdrop of a brilliant snowy scene illuminated by blazing buildings and the acrid smoke of burning fish and oil that smelled like a backstreet chip shop. The score card was kept: 216 German prisoners captured, plus 60 Norwegian quislings arrested; 315 Norwegian volunteers going to England; an armed enemy trawler captured and taken back to England with the loss of 14 German crew, killed or wounded; 11 ships sunk (22,000 tonnes); 18 factories blown up; and 7 fish, oil and petrol installations burned to the ground or blown up, with one English manager rescued.

The only British casualty was an officer in No. 4 Commando who carelessly stuffed his .45 Colt revolver in his pocket and shot himself in the thigh when he slipped on the frozen ground. The return was speedily achieved, and stories were exchanged about the raid on the journey home. There was the party who almost got shot up by the guns of one of their own ships when both arrived at the same target vessel at the same time. There were the soldiers searching an administrative building who burst in on two German officials who were still in bed asleep. The locked door flew open with one shot; a chamberpot, clearly visible under one of the beds, provided an irresistible target. And back on board, Provost-Sergeant Bill Chitty, herding the quislings to their accommodation, accidentally pressed the wrong button on his Tommy-gun (one of only two available for the raid) and loosed off a hail of bullets which landed between the feet of his CO, John Durnford-Slater. 'You must be more careful with those bloody things,' said the CO; 'they're dangerous.'

News of the raid gave true cause this time for one of those fulsome communiqués about which Churchill had complained last time when there was nothing much to crow about. Although there was no fighting to speak of, the significance of the raid became apparent later after two more assaults on Norway in 1941. One called for a return to the Lofoten Islands to capture a German garrison that had been set up since the first visit by the Commandos. A force of 300 men from No. 12 Commando made a forced landing and took the garrison by surprise, capturing and killing a number of German soldiers before retreating as quickly as they arrived 24 hours later. Soon afterwards, following another couple of false starts, a more ambitious assignment was allotted to No. 3 Commando, with additional support of two troops from No. 2 and No. 4 at the end of December 1941. The target this time was the strategically important Vaagso Island, at the entrance to Nordfjord, with targets at South Vaagso and Måløy island, where German garrison and oil installations had been established. The raid was expected to be the toughest yet for the Commandos, and so it proved.

After a horrendous crossing from Scapa Flow, with the transport ships knocked around by gales and heavy seas, the group this time had the support of the RAF, which were to drop smoke bombs to provide cover for the landings and to attack marauding Luftwaffe who, sure enough, came zooming out of the skies as the assault parties prepared to land. One landing craft took heavy casualties on the way in when it was hit by a phosphorous bomb from a British aircraft hit by German fire. The troops were split into five groups, each with specific targets. Installations to be attacked by the demolition team had been earmarked in advance, and the troops had to cover a good deal of ground to reach them. Major Jack Churchill, second in command of No. 3 Commando, led his men ashore in his usual trademark style playing his renowned bagpipes and then leading the charge swirling a claymore around his head. The action was at times fierce, and every one of the units saw heavy fighting, often resulting in house-to-house and then hand-to-hand battles, all played out against the background noise of explosions and flames as the demolition teams set about their business. An account in the logs of Combined Operations, under whose auspices the raid was set up, provides a taste of one of the encounters by troops from No. 3:

> We had to get through the streets and warehouses where snipers were doing their stuff and holding our troops up. The orders were to go through the town and make contact with some other troops. Our captain led the attack here, and although it was slow as we had to go from house to house, we were able to spot and shoot the snipers who were doing the damage. One of our sergeants received three shots in the back from a sniper who had let us pass. We opened fire on this sniper's window and settled him. We dashed to the next house. There was a German here who was a tough fighter. When the captain put his head round the door, the German fired and so did the captain. Both missed, and we threw grenades, a lot of grenades. The Hun just came to the door and dropped two stick grenades near us. One of our officers, who had already been wounded through the shoulder, went in after him with a Tommy-gun. The Hun shot him, but he was able to walk out.
>
> One of the section dashed in and sprayed a magazine Tommy. The Hun shot him through the thigh and then let him get away. We decided to burn him out. We threw in petrol, set fire to it and went on our way, leaving one man to deal with the Hun when he eventually appeared.
>
> From here on we fought from house to house, with several

> casualties, until we finally reached the other end of the town. We then made contact with the other troops, and the whole town was cleared out and the battle was over.

The Commandos took the heaviest casualties at Vaagso since their formation: 17 killed and 53 wounded. They also took 98 prisoners and arrested four quislings. The attack was undoubtedly a success in every regard, but especially in one of its prime aims, which was to force Hitler to reinforce the German garrisons in Scandinavia and thus divert troops from eventual deployment in other parts of Europe. The idea, as Lord Lovat pointed out, was to create a scenario where Hitler and his planners had to keep a constant eye on the possibility of an Allied invasion of Norway. It worked. A week after the Vaagso raid, Hitler issued a directive in which he stated that 'Norway is the zone of destiny in this war. I demand unconditional obedience to my orders and directives concerning the defence of the area.' He sent a personal representative on a tour and, on his recommendation, three more divisional commands were established and many more gun emplacements were set up to cover the approaches to the fjords and harbours. As Lovat pointed out: 'These assaults forced German High Command, bowing before a Hitler directive, to reinforce the northern coastlines . . . but as it happened the Allies never returned to Scandinavia. On D-Day the Norwegian [German] garrison numbered 300,000 men. It is fascinating to consider the difference these divisions could have made in Normandy. Either way . . . [Hitler] made a mess of it.'

Chapter Four

Not According to Plan, Again

The fortunes of the Commando units based in the United Kingdom during 1941 were not matched by those of Layforce. It will be recalled that the combined manpower of Nos. 7, 8 and 11 Commandos and a troop from No. 3 was marshalled by Lieutenant-Colonel Bob Laycock and dispatched to the Middle East to link up with the 300,000-strong British army on the Nile under General Sir Archibald Wavell. In spite of his successes in the Western Desert and the capture of 100,000 Italian prisoners, Churchill was repeatedly critical of Wavell's apparent reluctance to attack on other fronts.

Wavell replied that he was already committed to the capture of the Italian Dodecanese Islands and planned to move on to Rhodes early in the New Year to forestall a German base being established there. In the meantime, the arrival of the German Afrika Korps, which landed at Tripoli on 14 February 1941 under the command of General Erwin Rommel, cast further doubts on whether the British forces could maintain their supremacy in the Middle East. Rommel's orders were to rescue Mussolini's forces, which had suffered a series of humiliating defeats, and to halt the British advance across Libya. The RAF, which effectively controlled the air war, would also be challenged with the dispatch of Luftwaffe squadrons to back up the Afrika Korps.

As will be seen, although it has been said that Wavell requested the presence of Layforce, this is now in some doubt, given the testimony of some of those present at the time. Certainly, he had been a supporter of Commando operations that had already proved to be effective in the region. But Wavell hadn't used them in the manner for which they were formed, i.e. as raiding parties. Had anyone sent him a copy of the Commando manual?

The Middle East Commando's initiation into action had been almost as frustrating as that of their counterparts in the UK. Three Commandos,

Nos. 50, 51 and 52, were formed, recruited largely from experienced regulars already serving in the desert. They had suffered a number of on–off false starts and were then variously deployed in prolonged campaigns in Sudan, Eritrea and Abyssinia as part of the attacking programme of larger formations, which was not what the Commandos were trained for; theirs, as ever, was allied to purely raiding missions. No. 50 Middle East Commando had also been posted on the island of Crete in a defensive role, although it was often hampered by lack of transport and suitable armoury.

Wavell can, however, be credited with one far-sighted innovation, by giving the go-ahead for the creation of a new force that ultimately became recognised as one of the earliest and most successful formations in the history of Special Forces: the Long Range Desert Group (LRGD). This was founded at the suggestion of former First World War sapper Major Arthur Bagnold, who was among the early pioneers of desert troop mobility. He proved his point in the late 1920s with the Royal Engineers by driving a model-T Ford into the great sea of sand at a time when the general consensus was that motorised vehicles could only be used on roads. Bagnold recalled:

> So [in June 1940] when the hot war broke out, I took my courage in my hands and wrote out a note which I asked a friend of mine to put on the Commander-in-Chief's personal desk. And within half an hour I was sent for. Wavell was alone . . . he sat in an armchair and he said, 'Tell me about this . . .'
>
> I suggested that we ought to have some mobile ground scouting force . . . able to penetrate the desert west of Egypt . . . I said, without thinking, some 'piracy on the high desert'.
>
> 'Can you be ready in six weeks?' Wavell asked.
>
> 'Yes . . . provided . . .'
>
> 'I know, I know . . . there'll be opposition and delay.'
>
> Wavell then rang his bell and a lieutenant-general came in. He dictated a message . . . 'I wish that any demand, any request made by Major Bagnold in person should be met instantly and without question.'
>
> It was typed and Wavell signed it.

The LRDG was on the road, and within the year Wavell had sent the following dispatch to London, after another series of successful raids:

> I should like to take this opportunity to bring to notice a small body of men who have for a year past done inconspicuous but invaluable

> service: the Long Range Desert Group. It was formed under Major (now Colonel) Bagnold to reconnoitre the great Libyan desert on the western border of Egypt and the Sudan. Operating in small independent columns, the group has penetrated into nearly every part of desert Libya, an area comparable in size with that of India. Not only have patrols brought back much information but have attacked enemy forts, captured personnel, transport and grounded aircraft as far as 800 miles inside hostile territory. They have protected Egypt and Sudan from any possibility of raids and have caused the enemy, in a lively apprehension of their activities, to tie up considerable forces in defence of distant outposts. Their journeys across vast regions of unexplored deserts have entailed the crossing of physical obstacles and the endurance of summer temperatures, both of which have been achieved only by careful organisation and a very high standard of enterprise.

Although not a recognised Commando outfit, the LRDG was the torchbearer of great desert journeys and long-distance raids hundreds of miles behind enemy lines. It was a model that David Stirling – soon to be in action in the region as a member of Layforce – would pick up and modify for his own SAS. In fact, the group was the only effective force operating behind enemy lines prior to the arrival of Layforce, around which there was great anticipation and high expectation from its men. Sadly, the whole shooting match for which they had been deployed was soon to collapse around their ears.

Layforce departed Scotland in a convoy of ships and under heavy escort on 31 January 1941, heading out into the North Atlantic to stay out of range of long-distance bombers. The full force of the Atlantic swell gave them all a bad time before the convoy veered south around the tedious Cape route to arrive in Alexandria on 11 March. James Sherwood, then a lance-corporal, was among them:

> We were given a rousing speech by Laycock before setting off . . . sailing straight into a force-9 gale, and it certainly felt like it in this 10,000-tonne ship carrying its landing craft slung from davits either side of what would normally be the boat decks, with the Commandos on board in hammocks slung in the former cargo holds, all very makeshift. They were very makeshift toilet facilities, too, merely long rows of seats with holes in them facing one another down a long alley and water running along underneath, as primitive as they come and all very matey. Everybody was so sick and unhappy in this heavy sea they couldn't have cared where they sat or who they faced or anything else in their misery.

It was a fairly fast convoy, although I suppose it took us about five weeks to get out there via the Cape, landing eventually at Suez, where we were immediately struck by the discomfort of millions of flies. Our base initially was Geneifa on the shores of the Great Bitter Lake, a transit area, really, of ramshackle tents, very, very basic equipment, no comfort whatsoever, no showers for washing and that sort of thing. Within a couple of days of getting there we were all smothered in a sandstorm which lasted about 24 hours, blew the tents down, buried us and our kit literally. All we did was keep our heads down under blankets.

We stayed there for about a week and did a few half-hearted route marches through sand, which wasn't very well received. Then two of us were teamed up with Roger Courtney and moved down to a naval base at Kabrit, at the other end of the lake. There we were joined by Lieutenant-Commander Nigel Willmott.

Nigel Clogstoun Willmott, a 30-year-old senior naval navigator for ships, was working on plans for a shipborne invasion of Rhodes. As a veteran of the Narvik disaster a year earlier, when the British had suffered heavy losses as they foundered on rocks and hidden shoals off the Norwegian coast, Willmott was well aware that many ships' navigators had in the not-too-distant past been civilians and at best had only amateur navigational experience. He put forward a strong case for a reconnaissance of the island. Laycock agreed, and teamed him up with Courtney. The two men were to be taken under cover of darkness to two miles or so off Rhodes aboard the submarine *Triumph*. They would then paddle away in canoes, equipped with sub-machine-guns, Tommy-guns, grenades and a Thermos of coffee laced with brandy, to recce possible landing sites on Rhodes for Layforce and the mainstream troops who would follow them ashore. James Sherwood was given the task of being the backstop for the two men:

After several nights of dummy runs and training, we set off, a journey lasting a couple of days to get to our point off Rhodes. It was very exciting because I'd never been in a submarine in my life. I'd always seen them as rather romantic things. I soon came to learn that it wasn't so. My role in this was not to go ashore but to keep the two canoes in good order and maintenance and be up on the casing of the submarine for the launching and return. When it came for them to set off, it was pitch dark. The technique they developed involved sitting in the canoe on the casing of the submarine which then slowly trimmed down, sank itself in the water, until they just floated off. It sounded simple, but it wasn't.

Courtney and Willmott slipped away under the moonless skies and paddled ashore to the beach areas that would make suitable landing sites for the Layforce, noting data such as depths and rocks in chinagraph crayon on a slate-board. Finally, Willmott went ashore, dodging enemy sentries, to make a map of the terrain and nearby roads. He penetrated to within 60 metres of a large Axis headquarters at the Hôtel des Roses, apparently crawling about the lawn to get an idea of its troop population. On the following night, he and Courtney made a recce of the main beach south of Rhodes town, with Willmott this time cutting through wire barricades to get on to the main highway.

A third night was spent making a beach recce through *Triumph*'s periscope, and on the fourth and final night they again set off for the shore. Courtney swam to one beach, leaving Willmott to travel a little further down the coast. Willmott was to return and pick up Courtney, who would signal his position by dimmed torchlight. Courtney, however, ran into triple trouble. He suffered severe cramp while swimming, and as he lay writhing on the beach he attracted the attention of a noisy dog, and then to cap it all his torch failed. Willmott managed to find him and brought him to safety only minutes before an enemy patrol appeared on the beach. Had Courtney been caught, he would undoubtedly have been shot. This was the first major beach reconnaissance of its kind. Both men were decorated for the mission – Willmott was awarded the Distinguished Service Order and Courtney the Military Cross – and promoted to captain. Their meticulous charting of potential assault beaches would later become one of the prime tasks of a most secret wartime organisation called Combined Operations Pilotage Parties, headed by Willmott himself, set up to guide major invasion forces in the latter stages of the war. COPPs, as they were called, was so secret that their very existence was not revealed until a dozen years after the war had ended.

In this instance, however, all the dangerous work carried out by Courtney and Willmott was to no avail. The Rhodes landing was cancelled even as Layforce prepared for the assault. Rommel's Afrika Korps had succeeded in driving Wavell behind the port of Tobruk. Meanwhile, Hitler had ordered the invasion of Greece, and the Germans were on the brink of taking Crete and Rhodes. Wavell began using Layforce in the way he had used the Middle East Commandos. Raids were consistently being set up and then cancelled; frustration grew to anger; confusion and then chaos reigned. Gradually the force began to be cut to ribbons by misuse, bad planning and sheer bad luck. What happened next shook the whole Commando organisation, even those back in Britain when they heard about it, and James Sherwood was still fuming half a century later:

> Layforce was soon to discover that in fact we weren't wanted – and discovered it directly from General Wavell himself, who didn't seem very keen on the idea of having a group of Commandos under his control. And it was put very politely. In effect, he said: 'I don't know why you've been sent out here.' This was a great shame because we had three full Commandos of men of really the best kind of spirit, all being pure volunteers, in it because they wanted some action. We were there when he came along and addressed us. We were completely taken aback by this, very disappointed, especially as we had already done the recce at Rhodes in anticipation of that operation going ahead. While we had been expecting a big welcome, here we had the Commander-in-Chief telling us he didn't know why we were there. At the same time, the Germans were literally knocking at the front door.
>
> The shocking outcome of all this was that Layforce was partly disbanded and scattered around the Mediterranean in penny packets. Some went up to Tobruk, some went to Crete. The No. 11 Commando suffered pretty considerable losses in action against the Vichy French, in south Lebanon, just over the border from what was then Palestine. So it was really a story of disbandment of Layforce, of the three Commandos, and of their reappearance in various forms, or return to unit for chaps who couldn't find employment in some other form. David Stirling, then unemployed, formed his SAS out of some of the remnants of Layforce, and Roger Courtney through sheer perseverance managed to carve a niche for the SBS.

Sherwood's nutshell account of the way in which things developed takes us ahead of the action. The curious reluctance to use Layforce on the kind of operations for which the men had been raised, in a scenario that called out for such activity, was apparent from the beginning. Rommel had launched new offensives, retaking much of the ground Wavell's army had won from the Italians. On 11 April Layforce was assigned the task of a detailed reconnaissance of the coastline of the Western Desert, sending raiding parties of 200 men ashore. No. 7 Commando set sail in HMS *Glengyle* but had barely got away when the orders were changed: they were to carry out harassing raids against the Germans at Bardia and Bomba. Then the Bomba raid was cancelled and they headed 250 miles down the coast to Bardia in a convoy consisting of the Commando ship, an escorting anti-aircraft cruiser and three Australian destroyers; it was indeed a very costly operation to set up, and Laycock was keen for his men to do well in this first raid.

As soon as they arrived in the target area for the raid on the night of

20 April, the plans started to go awry once more. Roger Courtney's folboat team were travelling in the submarine *Triumph* and should have arrived in advance to provide navigational markers. But the submarine was attacked by British planes, had to perform a detour and Courtney's canoes were damaged. Meanwhile, it took two hours to launch the landing craft from the *Glengyle* because of jammed mechanism, but surprisingly there was no enemy presence. Intelligence for the raid proved to be well wide of the mark from beginning to end. They were split into three groups, each with given targets. Most of the installations to be attacked simply did not exist at the marked locations or were worthless. In fact, the only tasks completed were the blowing of a bridge and the firing of an Italian tyre dump. Then the force had to withdraw because of the preset time-limits ashore to allow the convoy to retreat before daylight.

As the return got underway, an officer in one group was accidentally shot and mortally wounded by one of his own lookouts. One detachment arrived at the wrong beach for re-embarkation, and 67 men were captured. One of the landing craft broke down and the *Glengyle* sailed without it. The men eventually put to sea and managed to get ashore at Tobruk three days later. The Commandos' first raid was a comedy of errors, the largest of which was that they were sent to Bardia at all for such a brief period in which virtually nothing was achieved. Thereafter, and for some time to come, they went through a period of similar situations: working up for raids that were subsequently cancelled, or achieving little. As Charles Messenger recorded in his excellent account,[1] when the *Glengyle* was eventually handed over to assist in troop movements at the time of the fall of Greece and Crete, an inscription was found on one of the troop decks which read: 'Never in the whole history of human endeavour have so few been buggered about by so many.'

This message, said Messenger, was reasserted by Laycock himself at a lecture on his return from the Middle East: 'I think most of us imagined that conditions in the Middle East would be very different from the old practice of working up to fever pitch for operations that were eventually aborted, a thing of the past. I can assure you . . . that was not the case . . . [it was] if anything even more exasperating.'

The demise of Layforce as a working group under a single command structure began towards the end of April with the fall of Greece where the Allies were forced into retreat after ferocious battles that overwhelmed Australian and New Zealand forces. By then, the Enigma

[1] William Kimber, *The Commandos, 1940–45*, 1985.

decrypts from Bletchley Park revealed that the Germans were planning a massive invasion of the strategically important island of Crete.

Churchill ordered Wavell to send reinforcements and more guns, but the commander wired back that he could spare only six tanks, sixteen light tanks and eighteen anti-aircraft guns, although the Allied manpower on the island was bolstered to around 30,000 men after the fall of Greece, with British, Australian and New Zealand troops under the command of General Bernard Freyberg. Even so, Churchill, having viewed the Germans' precise order of battle – courtesy of the Enigma decrypts – clearly had doubts that they had sufficient firepower to hold on. Freyberg was desperately short of heavy equipment, much of it having been abandoned during the retreat from Greece. His worst fears were realised on the morning of 20 May when, as predicted by Enigma, German Stuka dive-bombers and artillery aircraft roared and screamed over the horizon and began pummelling the Allied troop positions. They were followed by wave after wave of stinging aircraft attacks and landings, including Ju-52s towing huge DFS-230 gliders packed with troops, vehicles and guns. Suddenly, the skies were filled with the greatest airborne invasion force ever mounted in the history of warfare.

By late afternoon, almost 5,000 men had been dropped or landed on the island, and one of the most costly battles of the war to date was underway as more German paras and mountain troops were delivered hour after hour, eventually totalling 22,040. They met spirited Allied resistance whose strength had been hugely underestimated by German intelligence. Even so, Freyberg was in dire trouble from the outset through his lack of heavy artillery and, the greatest weakness of all, virtually no air power to meet the Germans' massive aerial bombardment, which went on, unrelenting, for five days.

No. 7 Commando and 50 Middle East Commando, now known as A and D Battalions respectively, entered this maelstrom at around midnight on 26 May, led by Bob Laycock himself, with Major Freddie Graham as his brigade major and a very grumpy Royal Marine captain, Evelyn Waugh,[2] acting as Intelligence Officer. They had sailed from Alexandria aboard three destroyers heading for a landing site off Selino Castelli. Rough seas, however, made it impossible to power the boats to take the Commandos ashore, and the naval captain decided to return to Alexandria rather than get his ships shot up. The Commandos were delivered to Suda

[2] Sir Ronald Swayne, No. 1 Commando, said of Waugh: 'He was very brave. He had no fear at all, so I'm told, in Crete. But he was terribly unfit, and his awful fault as an officer was that he was contemptuous of his soldiers, one thing you must never be as an officer. And he was so disliked he was frightened that one of the soldiers would put a bullet in his back. He never ought to have been in the army.'

Bay two nights later aboard a minesweeper that dropped anchor as close as possible. Even before the manoeuvre had been completed, all kinds of small craft began hurtling from the shoreline towards the ship. A bedraggled naval officer came on board and blurted out: 'It's bloody chaos out there. The army's in full retreat. We're evacuating.'

'Well, thanks very bloody much,' said Freddie Graham, who happened to overhear the conversation. 'We're just about to go in.'

They were apparently still needed, and battle raged around them. As Freyberg noted in a letter to Churchill after the war: 'We were broken, but only after some of the heaviest fighting of the war. There are over 4,000 German graves in the Suda Bay area alone and, I believe, a similar number of our own, a lot of them blasted out with 500-pound bombs.'[3]

The Commandos were to cover the retreat towards the evacuation points, and they had to move rapidly and under constant air and mortar attacks to their positions. D Battalion was placed on a ridge six miles east of the landing site and next morning was moved twelve miles south, which was considered a more suitable spot to cover the withdrawal. Meanwhile, A Battalion took up fighting positions as infantry, and aided by two tanks of the 7th Royal Tank Regiment and a force of New Zealanders drove back a forceful advance by the Germans. They were themselves now in danger of being cut off from the departure point, a dozen miles behind them. A Battalion was scattered and Laycock began his own withdrawal, falling back progressively to battalion positions, though not before they had kept the route open for hundreds of Allied troops to make their escape.

Meanwhile, high on their lookout hill, D Battalion were soon to be confronted by mechanised German units that could be seen advancing in the distance with tanks trundling forth at a fast pace of knots. They had clearly spotted D Battalion's mirrored signals and heavy stuff began to rain down around them. They had no alternative but to pull back. As they did so, all hell broke loose. Stuka dive-bombers zoomed over the horizon and a star shell was fired over the British troops. Men were falling in all directions, while others were simply diving for cover. Then D Battalion was confronted with a full-frontal attack by two battalions of the Germans' 5th Mountain Company. The battle raged for an hour before the Germans pulled back to await reinforcements. Again, they were held off with assistance from the 2nd Battalion of the 8th Australian Infantry Battalion, with D Battalion suffering 18 casualties.

By then, the withdrawal was well underway and D Battalion themselves began to pull back towards the beach. Freyberg left on the

[3] Martin Gilbert, *Finest Hour*, p. 1,100.

afternoon of 29 May, leaving orders for a surrender the following night to save any further casualties. By then it was clear that the Royal Navy, which had taken heavy losses in men and ships, could not withstand the battering any longer. The surrender was dictated to the Commandos' brigade major, Freddie Graham, by General Weston, the last of the commanders to leave. He gave permission for Laycock, Graham and Waugh to get aboard the last boat out. In doing so, they had no option but to leave a large portion of their surviving force stranded on the island, with the thousands of others who would now go into German PoW camps for the ensuing four years. British troops watching the last boats sail away began smashing their weapons in disgust, shame and anger. The cost had been high: 4,000 Allied troops killed, 2,500 wounded and more than 11,000 taken prisoner.

Of the two whole battalions of Layforce which went to Crete, only 23 officers and 156 other ranks managed to get away.

Even as the remnants of the two battalions were sailing back, a third – C Battalion, originally No. 11 Commando – was in action in Palestine, sent by Wavell to join the defence of that region. Churchill had been warning that the Vichy French had given the Germans the use of their facilities in Syria and that a full-scale invasion could be expected. Wavell was too tied up with Crete to worry about Palestine, and C Battalion was left pretty much to its own devices.

The Commandos arrived aboard the *Glengyle* in seas that the captain decided were too rough for landings, in spite of protests from the men, who were prepared to take a chance and retain the element of surprise. They were not allowed to do so, and consequently the enemy got wind of their arrival and were waiting. A series of running battles developed as the troops came ashore to head for their positions in support of the 7th Australian Division, which was fighting French Algerian units. The coastal route to the advance on Beirut was heavily defended, especially at the Litani River. C Battalion's orders were to land, seize and hold all crossing to the river, to be used by the advancing Allied force. But even before they arrived, a crucial bridge that would have carried them onwards was blown up. C Battalion spent a couple of nights surrounded by vastly overpowering enemy troops, who seemed to block every route forwards.

By the end of the first night, the battalion had lost a quarter of its men. They were relieved to discover that Australian sappers managed to build a pontoon bridge the following day, and with the advance on again a party of C Battalion's men who had been captured the night before now found themselves in the reverse situation – their captors surrendered to them. Even so, the battalion had been hard hit and was withdrawn to

Cyprus to perform garrison duties, which pleased none of them.

Even more of a shock was the news that Layforce was finally to be disbanded as a collective at the beginning of August because Wavell thought they were too costly and because the Royal Navy was becoming increasingly reluctant to get involved in amphibious landing operations because of the number of losses sustained by carrier ships. But having just performed the *coup de grâce* on the Commandos, Wavell himself was moved to India and was replaced by General Sir Claude Auchinleck, who proved to be a saviour to at least some of the unemployed Commandos.

At the time, David Stirling was in hospital with a back injury after his very first parachute jump ended in drama when his canopy failed to open properly and he came down to earth with a bump. While there, he mused over the plight of Layforce, in which he was still a lieutenant, and looked at the possibility of combining the tactics of the Commandos with the Long Range Desert Group and Roger Courtney's Special Boat Section, which was now running seaborne sabotage and clandestine reconnaissance missions behind enemy lines. Throw in that other newly acquired talent in the British military – parachuting – and there were the makings of a new special force. He envisaged something even grander and used the time lying on his back to set down his proposals, which he presented to Auchinleck as soon as he had recovered. The Commander-in-Chief liked the sound of it and so did Winston Churchill, who needed little persuading when he visited North Africa and asked to see Stirling personally.

The result was a gathering of like-minded young men – including the renowned tactical genius Captain Paddy Mayne, an Irish rugby international, and Lieutenant Jock Lewes, who just happened to be around when a large number of parachutes bound for India fell off the back of a lorry. Together, they formed a force that could drop out of planes, dash around in trucks, tanks and what have you, charge ahead of the herd and cause mayhem. The men were Commandos by another name and with greater versatility. Thus, the SAS was born and given a base at Kabrit, near the Suez Canal, in July 1941. Initially, it went under the grandiose title of L Detachment, Special Air Service Brigade, clearly hoping to fool the Germans into believing that the Allies had a new airborne brigade.

The concept of Layforce, on the other hand, hampered by lack of equipment, cooperation and enthusiasm from the top, lay in ruins, although it was not dead yet. Churchill would see to that. Laycock himself returned to England in July, basically to complain about the way his force had been used and about the general lack of cooperation he had received in achieving the original aims. He more or less pleaded with the

War Office for a more positive role and for it to revitalise Layforce rather than disband it. Winston Churchill, undoubtedly kept informed by his son Randolph, seemed to agree, and in a minute to General Ismay on 13 July 1941, wrote: 'I wish the Commandos in the Middle East to be reconstituted as soon as possible . . . the Middle East Command have indeed maltreated and thrown away this valuable force.'

When Laycock returned, there wasn't much left of his original force to command. The men had either been killed, captured, sent back to their units or volunteered for others such as the Long Range Desert Group. Apart from L Detachment (the SAS under Stirling), there was an HQ and depot troop at Geneifa, No. 3 Troop, made up of the remnants of Layforce, with a mere 53 other ranks left from the 1,500 who had sailed from Glasgow full of hope and bravado ten months earlier 4 and 5; Troops, made up of 51 Commando, and No. 6 Troop, which was Roger Courtney's SBS.

All these changes began to emerge as Auchinleck planned the long-delayed Operation Crusader, the new offensive by the 8th Army to retrieve ground lost to Rommel which Churchill had been pressing for since August and which was now due to be launched on the night of 17–18 November 1941. The aim of the offensive was to relieve Tobruk and recapture Cyrenaica. The Commandos were to get some ambitious side action: a daring – some said wild, mad and plain daft – scheme to infiltrate deep behind enemy lines to cause maximum commotion on the eve of the launch of Crusader by:

1) assassinating Rommel himself at his villa, supposedly in the village of Beda Littoria 190 miles inside enemy territory;

2) hitting the German HQ in the same place, and blowing up all telegraph and telephone installations in the area;

3) blowing up the Italian HQ at Cirene and the Italian Intelligence Centre at Appollonia; and

4) launching a secondary diversionary raid on the same night, with David Stirling's L Detachment attacking five enemy airstrips for forward bombers and fighters between Gazala and Tmimi – an historical event as the first ever raid of the fledgling SAS.

Bob Laycock was to coordinate the operations, although he was more concerned with the Commando raid, by far the more dangerous. This was under the command of Lieutenant-Colonel Geoffrey Keyes, whose father, Admiral Sir Roger Keyes, was head of Combined Operations in London. They were to travel by submarines to a beach landing area that had been surveyed by an SBS team comprising Lieutenant Ingles and Corporal Severn. They were taken to the area aboard the submarine *Torbay* a couple of nights before the raid, paddled ashore by canoe and made a complete recce of the beach. They also met British intelligence officer Captain Jock Haselden disguised as an Arab; he was to give the signal for the Commandos to come ashore on the night of 15 November when the raid would get underway. The party set off after nightfall aboard the submarines, the *Talisman* and the *Torbay*. The weather had just taken a turn for the worse, but the green light remained on for a landing in spite of rough sea, with SBS men guiding the heavily laden troops off the submarine casement into their tiny craft. As forecast by the SBS men, the launching of canoes and dinghies from the submarines would take for ever in those conditions. The *Torbay* carried the bulk of the raiding party, with 36 men aboard. It took almost seven hours to land them instead of the ninety minutes estimated. By then it was deemed impossible to land the remaining eighteen men aboard the *Talisman*, although Laycock, who was on the *Talisman*, eventually made it ashore with six others. He decided to remain at the rendezvous with reserves of ammunition, leaving Keyes, the group commander, to lead the raid itself.

There was one other problem: some confusion over the location of the house used by Rommel. It was now thought to be at Sidi Rafa and not Beda Littoria, the designated target. Two Arabs who came with Haselden confirmed this, forcing Laycock to switch the main attack. The depleted group set off on the 18-mile march to the inland target in driving rain and thunderstorms. Every one of them was already drenched from the landing debacle, and the passage through thick mud and slippery rocks as the march proceeded only added to their woes. At dawn, and making slow progress, Keyes ordered the men to take shelter in a cave, where they laid up for the rest of the day and began their move towards the targets at nightfall, reaching them on the night of 16–17 November.

They arrived at what they thought was the Rommel house at around 2330. Keyes placed his men in key positions: three were detailed to put the electricity plant out of action, five were posted around the building to cover all exits, two others were placed outside a nearby hotel to prevent anyone leaving and others covered the road on either side of Rommel's house. All were in position just before midnight when a recce was made of the house. Unable to find a way in through the rear, as planned, Geoffrey Keyes and his party walked boldly to the hefty front door and beat on it. Captain Campbell, who spoke fluent German, demanded entry and eventually a sentry opened the door. He was set on immediately by Keyes, but the German grasped the muzzle of his Tommy-gun and backed against a wall. Keyes was unable to draw his fighting knife, and neither Campbell nor Sergeant Terry could get close enough to stab the German in the throat in the manner in which they had been trained.

Finding it impossible to dispose of the sentry silently, Campbell shot him with his .38 revolver.

The noise alerted others. Two men came down the stairs in front of them, saw the commotion and were met by a burst of fire from Terry's Tommy-gun. Off the main hall were several doors. Keyes gingerly opened the first, but the room was empty. Inside the second room, from which no one had attempted to emerge, were about ten German soldiers in steel helmets. Keyes opened the door, saw them, fired a few rounds from his revolver and slammed the door shut. Campbell told him he would throw a grenade and pulled the pin. As Keyes opened the door half a dozen guns blazed and Keyes fell mortally wounded just as Campbell hurled his grenade. It exploded, killing all inside who had not been finished off by Keyes' gunfire. Terry and Campbell carried Keyes outside, and he died within a few minutes.

Campbell went back into the house and to the side of it to round up his men and was hit by a bullet from a jumpy Commando who thought he was a German; it broke his lower leg. Campbell was brought to safety, and as he lay in pain outside he ordered the men to hurl all their remaining grenades through the windows of the building and to withdraw immediately. Realising the burden he would be to the men as they made their escape, he ordered them to abandon him.

Only 18 men made it back to the rendezvous, where Colonel Laycock had been waiting, passing the hours by reading a damp copy of *The Wind in the Willows* which he had taken from the submarine. They all then moved towards the beach, where they had hidden their boats, and signalled out to sea. Miraculously, the submarine *Talisman* rose up and headed closer to the shore, but by then Laycock's party discovered that their boats had gone; not a single one remained. Even if they were there, the sea was far too rough to launch them. Lieutenant (later Major) Tommy Langton, then with the SBS, described the sequence of events:

> We were relieved to see the arranged signal from the beach, but it was much too rough to launch a folboat. The [submarine] captain decided to send Lieutenant Ingles and Corporal Severn on a spare rubber craft with food and water. This was attempted, but the boat was washed adrift by the swell before the crew could board it. Later, the party ashore signalled they had found the boat. They did not know what had happened to the rubber boats which had been left on the beaches. The submarine captain suggested in his signal to the shore that they should attempt at dawn to swim out to the submarine, which was hovering 800 metres from the beach. There

> were, however, a number of men who could not swim anything like that distance, and others who could not swim at all.

The suggestion was declined on the basis of all or none. The submarine was in a risky position even under darkness, and with no apparent way of rescuing the men the captain decided to put to sea and signalled he would return after dark the following day. Langton continued:

> We put to sea again . . . and closed the beach very soon after dark the next night. The sea was considerably calmer . . . but we were dismayed to see no signals from the beach this time, so after waiting some time the captain decided to send myself and Corporal Freeberry in to reconnoitre. The beach was deserted . . . [then] we spotted a light which appeared on a hillside. It was the correct colour but not giving the correct [recognition] signal, so I was suspicious of it. We walked a little further and thought we saw a movement. We both heard a shout soon afterwards but found nothing, and, since we were by then some distance from our boat and liable to be cut off, I decided to return to it and wait.

Langton and Freeberry waited for several minutes but saw nothing further and decided to launch their folboat to paddle along the shore towards the location of the light. Langton flashed his torch, heard a shout but saw no signal in return. They beached again and were upturned as they did so, losing a paddle. Then Langton spotted the glow of a lighted cigarette in the undergrowth and realised the people ashore were the enemy. They clambered back into the folboat and headed back to the submarine, a feat completed with one paddle only through the brute strength of the 16-stone Corporal Freeberry.

Only later did the full story emerge. The survivors of the raid had been attacked by Germans; some were killed, others were captured as they tried to reach British lines and at least five were known to have been murdered by Arabs. Only two made it back to base in Alexandria: Lieutenant-Colonel Laycock and Sergeant Terry. They reached the British lines on Christmas Day after an incredible 41-day trek on foot through hostile countryside and desert. Jack Terry's first words were supposedly, 'Thank God, I'll never have to hear another word about Mr Toad' – a reference to the fact that Laycock had relieved the boredom by reading aloud from *The W in the W*.

Geoffrey Keyes was posthumously awarded a Victoria Cross for his bravery and leadership. The raid had been a costly failure, through no fault of the Commandos. Intelligence had been dismal, and the excursion also demonstrated that there was still much work to be done to bring the Commando raiders to a level of operating efficiency, although it had to be admitted by even their most ardent critics that the prevailing conditions and other pressures outside their control prevented any real work-up for such an important mission. The men were let down, as became apparent later when it was discovered that the value of the target buildings in the raid had been greatly exaggerated. The place turned out to be nothing more than a supply depot and, furthermore, the house attacked was never used by Rommel, although he had at one time used the villa that had been the original target at Beda Littoria. The greatest irony of all was that Rommel wasn't even in North Africa at the time. He was in Rome, taking a weekend break.

The second diversionary raid scheduled for the night of the launch of Crusader was no more successful, although more survived. The bad weather, with high winds and sandstorms, was hardly conducive to a successful parachute-landing, but after consulting his men Stirling decided they were going anyhow. The detachment was divided into five groups, to drop from bomber aircraft close to the target airstrips and blow them apart. Then they would make their escape and be picked up by Long Range Desert Group transport.

Each group had 60 incendiary and explosive bombs for the attack. Not surprisingly, when the aircraft came in over the designated dropping zone, they had to take evasive action to avoid enemy flak, and, with winds gusting at 45mph as the men leaped out, they landed miles off target out in the desert; a number of them were hurt. It was almost 10 days before the survivors began wandering back, 22 of the 60 who set out making it to a patrol position, the task half done. Stirling was furious with himself, vowing that such stupidity would never happen again. Next time, he promised, not a detail would be overlooked. The SAS was on the move, but for Layforce it was definitely all over, as indeed were the Middle East Commandos as a whole. They had suffered from the overall pressures of the worsening situation in North Africa and the Mediterranean. They had been shunted around from one operation to another and had few friends in the hierarchy who would take them on and use them for what they were; indeed, there was a general lack of appreciation of the Commando ethos among their commanders.

After the Rommel raid disaster, enthusiasm declined further. Although the title of Middle East Commandos was retained to please Churchill, the men were largely absorbed into general operations as the battle against Rommel's Afrika Korps became even more pressing. David Stirling, meanwhile, was expanding the SAS, utilising many of the remaining Commandos who had come out with Layforce. When Churchill made a brief visit to the Middle East, Stirling campaigned to have all special operations in the region – apart from those of the Long Range Desert Group – brought under his control. Churchill liked his style and the plot, and agreed.

This included the incorporation of the SBS contingent still in the Middle East, much to the chagrin of several of its members. Stirling renamed it the Special Boat Squadron (SAS) and placed Lieutenant (Lord) George Jellicoe, ex-No. 8 Commando, in charge. It went on to record some significant and daring action throughout the Middle East and Mediterranean, while its original founder, Roger Courtney, returned to England to form SBS Mk II in time for the Operation Torch landings in North Africa already being planned for the autumn of 1942. The SAS as a whole, meanwhile, began its own journey into history, one that continued long after Stirling was betrayed and captured the following year and spent the rest of the war in Colditz.

Bob Laycock, meanwhile, was promoted to brigadier and given charge of the Special Service Brigade, Middle East Command, to succeed Charles Haydon, who had been recalled to London. Changes were afoot in the whole setup. Roger Keyes, by then 67 years old, was a tired man and now sadly mourning the death of his son Geoffrey. He had nurtured Combined Operations into being and had led from the front but was now being replaced, and Haydon was sent for to become deputy to the new man at the top.

Chapter Five

The Exploding Trojan Horse

Lord Louis Mountbatten swept into the Richmond Terrace headquarters of Combined Operations, his trenchcoat swirling and his vanity positively glowing, enhanced by the knowledge that his exploits in the Mediterranean were soon to be immortalised in the 1942 movie *In Which We Serve* by his very close friend, Noël Coward. He was the hero of the hour, in spite of his unfortunate experience of having had his ship, HMS *Kelly*, in dock on three occasions with substantial damage. Two of them could only be described as resulting from unforced errors which even his biographer Philip Ziegler admitted made his ship 'the laughing stock of the Navy'.

He arrived to succeed Roger Keyes as Director of Combined Operations in the autumn of 1941 with great gusto and a touch of overambition. Up his sleeve were lots of schemes, some good, some bad and some downright foolhardy. But he was going to be good news for the Commandos in regard to pushing their cause because he knew that was what Churchill wanted. Indeed, he was to become the sponsor and creator of a number of new groups whose *modus operandi* was based on the Commando ethos of black-faced daring. True, a fair number would get killed, but he explained away that, too, in the words he wrote for Noël Coward's film script about the fictional HMS *Torrin* (which was in reality his own HMS *Kelly*, which was finally sunk off Crete in May as the Commandos were themselves helping the evacuation):

> The *Torrin* has been in one scrap after another, but, even when we've had men killed, the majority survived and brought the old ship back. Now she lies in 1,500 fathoms and with her more than half our shipmates. If they had to die, what a grand way to go! And now they lie all together with the ship we loved and they're in very

> good company. We've lost her, but they're still with her, and we'll all take up the battle with even stronger heart. Each of us knows twice as much about fighting and each of us has twice as good a reason to fight. You will be all sent to replace men who've been killed in other ships, and the next time when you're in action remember *Torrin*.

Well, it was a real crowd-puller at the time, great for morale and unbeatable propaganda, even if a large number of widows and orphans did not concur. Coward received an Oscar for the film and the *New York Herald Tribune* said, 'Never has there been a reconstruction of human experience which could touch the savage grandeur and compassion of this picture.' That was before some of the later dramas of the war began to unfold.

Mountbatten was Churchill's choice to succeed Keyes, although there were a number of detractors, especially among the naval hierarchy. To push the point further, however, Churchill let it be known in the spring of 1942 that Mountbatten would be promoted to Acting Vice Admiral, his title would be changed to that of Chief of Combined Operations and as such he would be given 'full and equal membership of the Chief of Staffs Committee', which was something of a slap in the face for his opponents. The First Sea Lord, Sir Dudley Pound, was so concerned as to the wisdom of this appointment that he wrote to Churchill to express his concerns that it might be seen that a) it was being done on his advice, which it wasn't, and he could 'not shoulder the responsibility for it'; b) it was made contrary to his advice, which would give the impression that the Prime Minister had overridden his views; or c) the appointment was made because Mountbatten was royalty, which would do him harm in the service. In any event, said Pound, the Royal Navy would not understand 'a junior captain in a shore appointment being given three steps in rank'.

Churchill was not deterred by the opposition and explained: 'I want him to exercise his influence upon the war as a whole, upon future planning in the broadest sense, upon the concert of the three Arms . . . Combined Operations in the largest sense not only those specific operations which his own organisation will execute.'[1] As an afterthought, Churchill added that he was prepared to proceed on a step-by-step basis 'to see how we get on' and, knowing Mountbatten's penchant for PR, decreed 'no publicity at present' to the changes.

Mountbatten sought to mix Commando-style operations with some of

[1] Martin Gilbert, *Road to Victory,* pp. 71–2.

his own developing theories. He loved the idea of the Special Boat Section and Courtney's canoe-borne saboteurs. He envisaged an expansion on all fronts, amalgamating the theories of Commando-style raids with all possible forms of amphibious assaults, from two-man teams in canoes to major landings of men and machines.

At that time, just to recap, the UK-based formations consisted of Nos. 1, 3, 4, 5, 9 and 12 Commandos. Others would soon be added along with emerging units then being formed by the Royal Marines and the Royal Navy, whose own Commando units were finally to be launched after early frustrations through lack of funds, equipment and manpower. Mountbatten encouraged ideas and would usually give a hearing to schemes that others would have simply chucked out of the window. He would listen to them all and, according to Ziegler, 'the more outrageous the methods, the more he relished them'.

Mountbatten was just what the Commandos needed, a man of unswerving ambition and sufficient influence to put them back on track and present them with new challenges and a real role. They had been stagnating; of that there was no doubt, with many operations cancelled almost at the point of the men going in. Training was still haphazard and left to individual Commando units. Although tougher than normal military units, training was relatively undemanding when set against the tasks that the Commandos were supposed to be carrying out. Because of the lacklustre air that existed within the Commando setup, some would even forgive Mountbatten for what many considered to be an irresponsible streak that would lead him to placing his men in situations in which the odds of survival were stacked against them from the outset, a criticism he overcame by the statement 'they are all willing volunteers'. And they were. He viewed their deployment across a much broader front than his predecessor who, it must be said, had been severely limited by the lack of support from commanders of the three services who were too busy with their own problems to worry about his.

Mountbatten knew it would be difficult to alter the views of those who felt there was no need for guerrilla tactics and that he, personally, was an unwelcome interloper. But he had sufficient support from the top, i.e. Churchill, to more or less get what he wanted in terms of setting the Commandos on a more determined course. He teamed a collection of eminent scientists and engineers with thrusting young officers of all three services, who came together in the newly formed Combined Operations Development Centre. Their aim was to dream up weird and wonderful schemes, inventions and devices that might be used against the enemy. Mountbatten knew, from his private conversations with Churchill, that the Prime Minister wanted assaults and sabotage aplenty around the European coastline, and there was no shortage of ideas in that direction. He also possessed this very large band of volunteers willing to carry them out – men just waiting to hit Europe. The Commandos were raring to go and too often had suffered the anguish of going into training for a raid and then having it called off at the last minute. As Ronald Swayne explained:

> It was rather like an athlete training for a race. You get very fit and you come to your peak and just at the time of the race it's cancelled and you go back and start again. We used to train for these raids and then they'd be cancelled. We were in it to fight, and I don't think any of us really wanted to kill Germans particularly, but one wanted to do what one was trained for. I was on two cancelled raids; some people were on half a dozen.

The Vaagso raid, described earlier, was the first in Mountbatten's tenure after he arrived at Combined Ops HQ, although he had inherited the plan from Keyes. He also received some honest and straight talking from Commando chiefs, including his new deputy, Charles Haydon, and was told that if these men were to be sent on their clandestine missions to confront the enemy in guerrilla-type situations, they had to have much more training than had been possible so far. This was not merely to increase the chances of a successful mission, but also to improve the chances of survival, which, as Haydon well knew, had been badly lacking in the Middle East.

To this end, the Commando Depot – later to be called the Commando Basic Training Centre – was opened in February 1942 at Achnacarry Castle under the command of Lieutenant-Colonel Charles Vaughan, CO of No. 4 Commando.[2] The castle was a partially burned-out baronial pile on the country estate of the Chief of Clan Cameron of Locheil, 14 miles from Fort William, Inverness, to which many a student would now be enrolled after a long and arduous train journey from wherever his Commando was based. It was set amid glorious Highland scenery in the shadow of Ben Nevis with plenty of rough terrain and impossible hazards on which to bring the troops to full fitness and skill. There were numerous mountainous landscapes close at hand for cliff-climbing and mountaineering courses and plenty of water for various forms of swimming experience, from naked in ice-cold water to full pack survival techniques, so that never again would the Commando boss hear the cry of 'But I can't swim!' or 'Do we really have to climb up there?'

[2] Lord Lovat took over as commanding officer of No. 4.

The courses devised by tough Achnacarry tutors were naturally all that might be expected to bring the men to the realisation that if they failed these, they might not survive the war. They began with basics such as unarmed combat, judo in three easy lessons and speed-marches over 15 miles of backbreaking terrain bearing full packs. There was the Death Slide, or how to cross a river without a boat: one man swims across with a rope or wire tied around his waist; one end is fixed 20 feet up a tall tree, the other tied to a lower trunk on the other side. There were exercises in boatcraft, with eight men packed into river assault craft known as Goatley boats with their weapons piled into the bottom. The paddlers sat on wooden slats, and if anything other than their arms were visible over the canvas sides, a volley of live machine-gun fire from their tutors on the shores hit the water 'unhealthily close to our craft', as one diary-writer noted.

Even tougher tactics were employed for exercises in opposed landings. Groups would need to dash through a quagmire, in which feet and legs sank if they lingered too long, and attack a mound, dive to the ground and fire at metal plates. Then they would slither and slide, shout and scream as they stormed fixed positions with live rounds from defending Bren-gunners cutting up the earth around their approach. Accidental shootings and other mishaps were not uncommon, occasionally with fatal consequences.

As the course developed, the exercises and training intensified. Elsewhere around the British coastline there was specific training in beach landings when each Commando took out a section for demolition and explosives training. As the Mountbatten–Haydon era became established, it was clear that instructions had been received from on high that any future Commando operations would be prefixed with clear, precise and specifically designed work-ups and drills; tutors fresh from nasty situations themselves dreamed up all kinds of horrors to confront the trainees. Ernie Chappell, still with No. 1 Commando, was in the thick of it:

> We enjoyed it, and I think those who stayed the course were a special breed of person. We were actually out to enjoy it. Most of us were by choice 'hostilities only' troops. Many soldiers weren't cut out for this sort of thing – those who had been schooled in regimentation rather than guerrilla tactics. They found route marches a chore. We considered route marches and training around the countryside an adventure. It was great fun. It was instilled into us that we were probably considered to be the finest fighting troops in Europe, and we believed this. This is why we were in the

> Commando. We liked it, we wanted it. We were belligerent. And we wanted to have a go.

The next opportunity was already being planned. During the winter of 1941–42 Chappell's training routine began to go in a particular direction, and it wasn't hard to pick up the threads that something was on the cards. No one ever asked what or where; they just waited for something to happen. This looked like a big one:

> I had already become proficient in explosives, learned how to demolish all sorts of funny things: blow holes through walls of houses so that you could crawl through from room to room, blow up houses, bridges, railway lines, telegraph poles, anything to do with communications. But suddenly our attention was directed to dockyards, and we were taken to examine pump-houses, the mechanism of lock gates: closing and opening them, where to place the charges to get the best effect to terminally damage the mechanism and so on. We also looked at all the ways you could silently sabotage or disrupt a dock, even if only partially, by destroying the electrics, the switchboards, mixing up all the keys to the gates or throwing all the keys in the water. Next, I was sent to a place called Burntisland, met with a motley crowd of Commando people from all over, for this supposedly advanced demolitions course. And for the next fortnight or so we spent our time examining the installations of the various docks in the Edinburgh area and at Leith and Rosyth. We talked to the employees on the docks. We examined the installations and saw how we would set about placing charges. Then we were split into a couple of parties; my group went to Cardiff, and the other to Southampton. I went to Cardiff, and we practised laying dummy charges. We became very proficient.

Chappell and his mates were preparing for something but knew not what, although they could more or less piece together the possibilities: they surmised that they were most likely to return to France and that, in the event, proved correct. Mountbatten was particularly interested in raids to damage the facilities for German ships along the Atlantic coastline of France, where deep-water ports and safe estuary docks were giving sustenance to enemy shipping of every description. There were five U-boat bases alone. Several were large enough to service German battleships and destroyers attacking Britain's crucial supply line to America. One port in particular presented itself as an early target for Mountbatten's new improved Combined Ops: St-Nazaire, at the mouth of the River Loire, which had apparently been earmarked by the

Germans as a dry dock and base for their most feared battleship, the *Tirpitz*. That port was to be the Commandos' target. According to Ronald Swayne, the idea for the St-Nazaire raid came from two Commando officers, Bill Pritchard and Bob Montgomery, who escaped back to England through the port in 1940.

They put their scheme directly to Mountbatten, who then had his own team work out the plot. By coincidence, Mountbatten had received reports of two remarkable inventions by the Italian navy: human torpedoes, driven directly to their target by two-man crews wearing breathing apparatus and, even more unlikely, an exploding motorboat. An example of the latter was picked up by the British off Crete. It was a small, high-powered craft that could be delivered to within 50 miles of its target by ship or even dropped from the air. Its bows were packed with 225 kilogrammes of explosives. It would be driven at breakneck speed by a single operative who, once within range of the target ship or dock, would lock the steering mechanism and pull a lever that ejected him and an inflatable life raft into the sea. He would then try to make landfall or, in a worst-case scenario, surrender to Allied troops.

The exploding motorboat appealed to Mountbatten, who immediately ordered a British prototype to be built. But the vision of an exploding vessel suddenly grew into something far grander than a tiny motorboat. Why not a bigger vessel? A motor torpedo boat (MTB), for example. And Mountbatten said, 'Why not a ship . . . a big ship, an old destroyer, for example?' Hence Operation Chariot was born. It did indeed look, on the face of it, to be an outrageous idea, but slowly it developed into a definite proposal.

They would sail a ship packed to the gunwales with explosives into the hugely fortified dock at St-Nazaire. It would be timed to explode approximately ten hours after its arrival, thus giving time for other explosive charges to be set and blown by teams of Commando demolition squads. The Commandos would follow on behind the ship in a flotilla of small craft, dodging the flak and defensive gunfire, then scramble ashore and set a dozen other massive explosive charges. Their task was to ensure that the dry dock at St-Nazaire was blown apart by the ship, and that various installations, such as the pump-house used to empty the dry dock, the power station and fuel supply lines, were damaged beyond repair. Additionally, two torpedoes with delayed timing mechanisms would be fired from two of the boats into the locks and set to explode at around the same time as the ship.

That was the plan, and Mountbatten thought it was wonderful: as Ziegler said, the more outrageous the scheme, the better, as far as he was concerned.

The hunt was now on for a ship to be used as the exploding Trojan horse: Mountbatten called for suggestions and was presented with an old destroyer, HMS *Campbeltown*, formerly an American vessel known as the USS *Buchanan*. Its funnels were cut down to disguise it as a German torpedo boat. Ernie Chappell was among the volunteers skilled in demolitions drawn from each of the Commandos. They would be accompanied by a fighting force from No. 2 Commando, and in all there would be 242 Commandos on the raid. As usual, those who gave the briefing for the mission made it sound simple and straightforward, although they did not underplay the dangers. The aerial photographs and scale models of the quay at St-Nazaire and the system of lock gates leading up to the dry dock showed a static situation that is always incomparable with the real thing: 'We approach from here . . . make our entry here . . . the main explosion will be timed for such and such . . . exit at this point . . . sail home. Simple.' Now let's see what really happened from one who was there – Ernie Chappell, who guides us through the action, or at least to the point where he lost consciousness with his wounds:

> After a week of preliminary training, the explosives team went down to Falmouth for final training and were put aboard a ship, the *Princess Julia Charlotte*. We were told that we were going to use our acquired knowledge to destroy a French dock. They didn't tell us which one, but from that moment onwards we were confined to ship and were allowed no communication with people ashore. We continued with more specific training and focused particularly on landings. We were taken off Plymouth and Devonport and we had to carry our mock raids on these ports during the early hours of the morning. The port authorities were not aware of our coming. So we could expect a hostile reception if we were discovered. In fact, some of our chaps were caught and put in civilian clink, so in that respect the mock raid was not a great success.
>
> Next we were shown the explosive charges we were to use on the raid. They were elaborate pieces of equipment specially made up at an experimental station somewhere. Each member of the team was allotted a task. My job was to connect the fuses in the main charge across the top of the dock gate at wherever it was we were heading and detonate.

After the technical preliminaries, Chappell and the men were, as he put it, introduced to boats and the Royal Navy crews who would carry the parties across the Bay of Biscay to their destination. They were, to say the least, shocked at what they found. Instead of tough, heavy-duty landing craft launched from landing ships that they had been expecting, they discovered they were to make the entire journey in 16 Fairmile motor launches known as Eurekas which, as Ernie pointed out, were basically plywood boats that were hardly suited to the task. They didn't have the range, for one thing, and extra petrol tanks had to be fitted on the deck, which they identified as an immediate hazard in the event of attack. The reason they were being used, it was explained, was that they were not uncommon in that area, being used on antisubmarine sweeping operations, and it was hoped that if they were seen no one would become suspicious.

The Commandos carried out numerous exercises in landings and getting ashore. The naval crew also needed to practise other intricacies for which there was nothing in the instruction manuals, such as maintaining station while the men went ashore.

The flotilla of MLs, along with one motor gunboat and one motor torpedo boat under the overall naval command of Commander R. E. D. Ryder, RN, were to follow in a particular formation in front of and behind the *Campbeltown*, which was under the command of Lieutenant-Commander S. H. Beattie, RN. The ship carried six tonnes of explosives, which had been cemented into her bows. Two groups of the assault force and five demolition parties also travelled in the destroyer and would jump over her bows or clamber down scaling ladders on to the dock at St-Nazaire after she'd rammed it. The rest of the demolition teams in the motor launches would come ashore at given locations. For Ernie Chappell's party, it was a spot called the Old Mole.

With a great buzz of excitement, the raiders gathered for a service and Holy Communion to calm them down. Then they were fired up again by a rousing speech from Lieutenant-Colonel Charles Newman of No. 2 Commando, who was in command of the raid, before setting off from Falmouth with two hunt-class destroyers somewhere in the distance for protection. The night of 27 March was very calm and the convoy had a remarkably uneventful journey across the English Channel into the Bay of Biscay. Ernie Chappell takes up the story:

> We did come across some French fishing boats and had to assume that they had German agents on board or wireless contact. So one of our escorting destroyers was dispatched to pick up the crews and sink the boats. I think it was determined afterwards that one of the ships had wirelesses and had reported the sighting.
>
> We sailed on down into the Bay until at the appointed time our escort destroyers left us. We were now on our own, although we had

a submarine submerged off the mouth of the Loire to provide navigational aid. We sailed on course for the mouth of the Loire, which we entered some time after midnight. Our target lay some six miles up the river. We hoped to make the biggest part of the six miles anyway without being discovered. *Campbeltown*, which was in the van, flew the German flag, which was quite a legitimate ploy of war, providing she didn't fight under it. A signals expert on board the ship knew the entry signals into the port of St-Nazaire, and we hoped he could fool the Germans at least for some time by giving messages to them in their code and by firing coloured Very lights, which we knew the Germans used in that location.

In fact, we had a fairly uneventful trip up the Loire for three or four miles. Then, suddenly, from the St-Nazaire bank searchlights panned the river followed by a short burst of fire. *Campbeltown* answered this with signals that we were in fact a German force who had encountered action in the Bay of Biscay and were making our way into St-Nazaire to repair damage with casualties on board, and could they please meet us at the quay with ambulances. This seemed to pacify the Germans for a time, and we made progress up the river without problems for probably another quarter of an hour.

At that time, we were again challenged from the shore. *Campbeltown* came under heavy fire, and we saw her German flag come down and the White Ensign go up. We were glad when that happened; we wanted to fight under our own flag. Heavy fire immediately opened up from both banks and from various ships in the harbour. We were being fired on from all directions, the stuff coming at us was . . . like all the colours in the world . . . a wall of metal. You had to be frightened. *Campbeltown* sailed on, making her way up towards the caisson,[3] which she had to ram. According to our plan, it was timed for 1.30 a.m. She actually rammed at 1.32. So we were two minutes out after navigating down from Falmouth and running the gauntlet of the River Loire. Commander Beattie, in charge of the *Campbeltown*, stood on deck and piloted the ship without flinching, straight up into the caisson and all this time under terribly heavy fire, really heavy fire.

Our MLs were coming up in two columns astern of the destroyer. We were under the same fire. We were firing back. The MLs were equipped with two heavy automatic guns, one for'ard and one aft, Bren guns and twin Lewis guns, and an old-fashioned light machine-gun. We engaged the guns on the bank but we were very vulnerable

[3] The watertight structure at the entrance to the dry dock.

> because of the extra petrol tanks on deck. We were in the rear, and by the time we reached our landing area several of our boats were on fire and were burning in the river. This made it virtually impossible for us to reach our designated position. When we got to the Old Mole there was a motor launch there, well alight. The sea was ablaze with burning petrol. One of the most pitiful sights was to see fellows swimming in the burning water and having to pass them by. They were screaming for help. We shouted at them: 'We'll pick you up on the way back.' We knew damned well we couldn't do that. We didn't like it, but we had to press on. Several of the MLs took heavy casualties. We tried several times to reach our landing area but couldn't get in. Ronnie Swayne, our commander, implored the captain of our motor launch, Lieutenant Ian Henderson, a great fellow, to run his boat ashore so that we could jump on the mud and get in that way. He wouldn't sacrifice his boat, and I don't blame him for doing that.

Meanwhile, a ferocious battle was now raging at the point where the *Campbeltown* had struck the entrance to the dry dock. Coastal batteries of six-inch guns opened up as the demolition parties and the assault force tried to scramble ashore. Not least in everyone's mind was that somehow the cemented-in explosives – whose timers had been set to blow at midday – would go up while they were all still in the vicinity. The ship took two direct hits on her bows, one of which blew off her 12-gun, killing the crew. Many of the Commandos were also killed or badly wounded before they could even get off the ship. The raid commander, Charles Newman, managed to get ashore with his headquarters group. They were supposed to link up with a party led by RSM Moss, who travelled in an ML and had been given the task of securing the forming-up point for Newman's HQ. Moss's group failed to arrive, and it later transpired that their ML had been blown out of the water by the shore batteries and all aboard were killed. Those demolition parties who did get ashore raced to their given tasks, often in twos and threes, and under an incredible hail of fire from all directions managed to set charges at some of the key installations.

Although wounded, Lieutenant S. W. Chant from No. 5 Commando, for example, reached the pumping station with Lieutenant Hopkins and Sergeant Dockerill and climbed down into the structure to reach the vital mechanism located 12 metres below. Working fast, fingers trembling and each with twenty-seven kilogrammes of explosives, they connected charges to four main and two subsidiary pumps linked by two parallel connecting fuses. Hopkins now pulled the two percussion igniters attached to the main fuse, and, in their webbing equipment, they had to retrace their steps back up eight flights of stairs lit only by small torches before the whole lot blew. They made it with only seconds to spare; they threw themselves on the ground just 18 metres from the pump-house when their charges exploded with a force that threw off concrete blocks placed around it by the Germans to prevent bomb damage.

Similar expeditions were made by others. The winding station was next to blow, and then another gate at the dry dock and, of course, the *Campbeltown* was sitting there, waiting to unleash her own deadly cargo, about which the Germans had no knowledge.

It was time to withdraw. The chaos, with gunfire all around, was simply too heavy to allow all the designated targets to be hit. Men were running to the quay and attempting to clamber into whatever boats were available. Lieutenant Chant missed one of the MLs that was just pulling away; it was hit in midstream and blew up, killing several on board. Others managed to swim ashore and were captured. With no boats left for an escape, Chant joined the others, who began to fight their way out of the mêlée, hopefully to escape, but again they came under heavy fire and the lieutenant could not carry on.[4] The rest met up with Charles Newman's party, which now numbered about 70 in all, several of them wounded. They began to fight their way to what Newman described as 'a long walk into Spain', but at daylight they were surrounded and taken prisoner.

Those who had managed to escape by boat had no easy task either. Ernie Chappell, whose ML was the last but one in the flotilla, arrived to witness the chaos ahead:

> After several attempts to land, it became obvious that the situation was hopeless as far as we were concerned. Several of the teams had got there. But we couldn't make it. The men were terribly disappointed when we had to turn tail and run the gauntlet back down the Loire, where again we came under terrifically heavy fire passing the coastal batteries. They gave us a good pounding, but we managed to get out into the Bay of Biscay and we were able to assess the damage. Although at this stage we hadn't suffered many casualties, the boat was riddled with bullet holes and the rigging was hanging in shreds. But we were now intent on making our way back to Falmouth under our own steam.
>
> This was somewhere around three or four o'clock in the morning, still very dark, very calm. We left the action behind us but then,

[4] He was eventually taken prisoner and repatriated as a wounded soldier in 1943.

> looking back at St-Nazaire, we heard some tremendous explosions and we could see glows of fire in the sky. And we thought, well at least somebody's done something. That's the powerhouse gone; that's the pumping station gone. We could sort of place these things. And we were very, very annoyed and mad at ourselves that we hadn't been able to get ashore.

As they began preparing to cover as much of the journey back before daylight, the ML crew suddenly heard the noise of a ship. The Commandos were already loading magazines on the ship's guns and preparing themselves for an attack. The ML captain, Ian Henderson, told them: 'Be quiet, lads. We're surrounded by shipping. I can't discern. We don't know whether it's our ships coming out of St-Nazaire or whether it's German ships.' The ML crew turned off the noisy engines and drifted for a while until a searchlight panned the sea and fell directly on the ML and followed up with a burst of fire. Ernie Chappell:

> It was in fact a German minesweeper bearing down on us at full speed, with the bow silhouetted in her own searchlight. It was like watching a great carving knife come towards us, intent on ramming us. It caught us at an angle and ran alongside. We were tilted over but did not capsize, and then we were raked with gunfire as the ship went alongside us. Our ML caught alight and we suffered some pretty heavy casualties. Henderson himself was mortally wounded at that instant. He had a leg blown off. He died on the boat. I myself was hit about the legs at that time and found myself with my legs dangling over the gunwales, slipping through into the sea. I was pulled back by some colleagues on board – couldn't say who – and was rested against the stack of our boat.
>
> Sergeant Tom Durrant, who was above my head on a twin Lewis gun, was wounded but carried on blasting away. Shortly afterwards, a shell hit the smokestack above my head and shrapnel from it penetrated my tin helmet. I was hit in the head and to all intents and purposes was out of the action and slipped in and out of consciousness. I remember the German boat circling us and the German captain calling out in English: 'Have you had enough? Have you had enough?' And he was answered each time with a burst of fire from Tom Durrant. I'm sure that we thought that our end had come anyway, and we had to fight it out. There was no point in giving up. So we continued the fight. Eventually, however, the ship circled us and the captain shouted again: 'Have you had enough?' And with our ML on fire and running out of ammunition,

> Ronnie Swayne saw that the situation was hopeless and shouted back: 'Yes, OK.' The German shouted back that if we were very quick he would bring his stern against our bows and take off all those who could walk. And here I think Swayne showed just what a character he was. He said this wasn't good enough; he couldn't accept that. He either brought his ship alongside and took everybody off, or no go, we'd go down with our ship. So the German captain agreed. His name was Leutnant-Kapitän Fritz Paul.[5] We didn't know it then, but he turned out to be a real gentleman. At the time, from my experience of being wounded in the legs and head, I don't think facing a firing squad would pose much problem other than the mental anguish. I must have been blinded at that time and remained so for some days afterwards. I can remember Swayne coming to me and saying: 'What's your wife's name, Chappell?' And I suppose I told him Frances. And I said: 'Where are we?' And he said: 'We're back at the quay at St-Nazaire.' I can remember saying: 'Oh, we'll make our way home through Spain.' He said: 'Shut up, shut up, shut up.' I don't remember any more than that until waking up in hospital. Actually, it was a casino which was taken over and used to house us and serve as temporary accommodation. I was lying in line with umpteen others of our fellows who had been wounded. They were in various states of dress and undress and awaiting medical treatment. I remember I was worried because I'd lost my false teeth. I never got my teeth back, may I say. So I went for the remainder of my prisoner-of-war life with no teeth.

Chaos and confusion, meanwhile, surrounded the St-Nazaire docks. Gradually, the fighting died down and the British troops who did not make their escape were being rounded up for transportation to prison camps while the wounded were carried away to the makeshift hospital in the casino. But there remained, still, the main event . . .

The sight of the *Campbeltown* lodged in the gates to the dry dock was the centre of attention, the Germans unaware of its lethal six-tonne cargo of dynamite hidden in her bows. Just before midday, the fuses activated and the ship blew apart with an almighty explosion that rocked the port, destroyed the dock and killed 400 Germans, including many officials who had come to inspect the damage. In mid-afternoon there were two more explosions – from the delayed-action torpedoes fired by the first

[5] Ronnie Swayne and Fritz Paul became good friends after the war.

MLs and which had lain silently embedded beneath the surface in the harbour structure.

The damage was greater than even the planners had anticipated, and the dock remained out of action for almost ten years. For that reason, the raid was considered a great success, certainly the best yet of all Commando operations. The human cost, however, had been high. Only five of the sixteen motor launches that set out on the mission and the motor gunboat made it back to Falmouth.

Of the 242 Commandos engaged in the operation, 59 were killed and 112 captured, many badly wounded. The Royal Navy suffered even worse casualties among the motor launch crews and those aboard the *Campbeltown*. They lost 85 killed or missing, 106 captured and, again, a large number wounded. Those who were capable of walking were taken to a compound exclusively for St-Nazaire captives and were later put in a prison camp, still all together, which was unusual, and were eventually joined by others recovering from their injuries. Because they were isolated, they at first feared they might suffer some particular fate, but in fact they were reasonably well treated and some of the more badly injured were repatriated in 1943.[6]

Back in London, Mountbatten was jubilant at the 'great success' of the first major raid originating under his command of Combined Operations and organised great publicity for his heroes. They were also recognised by an unprecedented number of medals considering the size of the force: eighty-three awards in all, including five Victoria Crosses given to the Royal Navy commanders Beattie and Ryder and to the Commando commander Lieutenant-Colonel Newman, who all survived, and posthumously to Sergeant Tom Durrant[7] and Able Seaman Savage, both for manning guns on board the MLs until they dropped, virtually cut to pieces.

Even as the survivors returned home, other raids were already on the drawing board. Not least among them was one that, as one wag put it, made St-Nazaire look like a Sunday-school outing: Dieppe.

[6] The prisoners who remained vowed to form an association after the war, as did many in such groups. The first meeting of the St-Nazaire Society was at the Chez Auguste French restaurant in Soho, London, where everyone drank heartily and toasted fallen comrades.

[7] Ronald Swayne: 'Durrant was enormously badly wounded; he was wounded all over, and I nursed him for some time and gave him morphia before he died.'

Chapter Six

A Cover-up

Even before the St-Nazaire raid was launched, many similar proposals were reaching – or being sent from – the desk of Mountbatten. The whole business of Commando raiding under the Combined Operations umbrella had changed in the space of a month of his arrival. He demanded and got ideas and input from across the board, and they did not simply involve Commando units, although the principles of assault were much the same. One of the most successful and least costly in terms of casualties, for example, was by 120 men of the newly formed 1 Parachute Battalion on a German radar station at Bruneval. The paras brought back radar equipment that had been enabling the Germans to shoot British planes out of the sky – and some technical prisoners to interrogate. Quite a few Germans were also killed. Now, with the success of the Commandos' Operation Chariot, new impetus was directed towards continued incursions on to the French coastline and a number of proposals were up for discussion. Some were sheer madness, others suicidal, and indeed a common but unadmitted theme running through all was the potential for a high mortality rate.

Vice Admiral Sir Charles Norris recounted a meeting at Richmond Terrace in which another raid on the French coast was being discussed; the idea was to land tanks at the Atlantic port of Brest and drop depth charges in the dry dock. Norris, then a young member of Mountbatten's staff, joked loudly to a friend: 'Let's forget all about this nonsense and catch the 5.15 from St-Malo.' Mountbatten overheard the remark and was at once struck by the idea of using the French railway for sabotage purposes and called for a plan to hijack a train, packing it with explosives and running it into the harbour.

That particular idea never got off the drawing board, largely through pressure of other commitments on the horizon. Even so, there were plenty of other developments, such as the exploding motorboat project and another picked up from the Italian navy – human torpedoes. These craft were manned by two operatives wearing breathing apparatus and could be released from submarines at a distance and then driven silently to their target, delivering explosive charges and then making their escape. Their existence was discovered in December 1941 when two Italians were found clinging to an anchor buoy off Alexandria. They were picked up at 0330 on 19 December by British sailors and taken aboard the battleship HMS *Valiant*, which was at the time sheltering at the port with its sister ship, HMS *Queen Elizabeth*.

The Italians became very agitated and then finally confessed that they were part of a six-man team who had piloted three underwater craft through the torpedo nets protecting the ships and attached time-fused charges on the hulls of both the *Valiant* and the *Queen Elizabeth*. The alarm went up, and crews of both ships dashed around closing all watertight doors. But it was too late to stop the explosions: the *Valiant* was rocked by a blast that blew a large hole in her stern and soon afterwards, the *Queen Elizabeth* reared up from two explosions. Both ships were knocked out of the war for weeks while repairs were made, leaving the British with only two battleships.

Mountbatten's Combined Operations Development Centre was ordered to look at these inventions and schemes. One of its members, the speed ace Major Malcolm Campbell, remembered an earlier paper written by a 28-year-old marine, Major H. G. 'Blondie' Hasler, who among other things had ideas for motorised submersible canoes. Mountbatten decided to form a special unit utilising small craft for the defence of Britain's own ports and to attack enemy docks. He placed Hasler, already the holder of an OBE, Croix de Guerre and Mentioned in Dispatches from the Narvik operations, in charge. The exploding motorboat had been given the codename of Boom Patrol Boat and thus Hasler's group would be called the Royal Marines Boom Patrol Detachment, staffed by volunteers from Royal Marine Commando units which were about to make their appearance in the war. The submersible canoes also came into being, under the codename Sleeping Beauty, and were famously used in an Australian-originated attack on Japanese shipping at Singapore.[1] But it was another Hasler project that most appealed to the chief. It would entail 12 men

[1] Operation Rimau was carried out by 32 swimmer-canoeist Commandos known as Group X, who launched their mission from Australia. Encouraged by Mountbatten, they were to use 15 submersible canoes to attach limpets to Japanese ships, but they then ran into trouble and heavy fighting ensued. Some were killed or drowned; eleven were taken prisoner, one died in captivity and the remaining ten were beheaded by the Japanese a month before the end of the war in the Far East.

being taken by submarine to the mouth of the Gironde, paddling by canoe 62 miles upstream against the flow to Bordeaux, which they would blow up enemy shipping. It was sheer bravado, and Mountbatten loved the idea. But the operation was considered by some to be too risky, and in any event it had to be put off until later in the year in view of other raids already in the pipeline.

The reason was soon apparent: three more raiding operations were planned for April 1942, hard on the heels of Chariot, the consequences of which were not entirely those envisaged. The first, which had a most disturbing climax, started out as an experimental excursion based on a raid by just two men, Captain Gerald Montanaro and Trooper Freddie Preece, of the Royal Engineers 101 Troop, a troop that used canoes and was later merged with the SBS. They were ferried under a moonless sky to a point two miles off Boulogne in an MTB. From there, they paddled into the harbour, conducted general reconnaissance and then attached limpet mines to a German ore carrier with 5,500 tonnes of copper on board, which was known to have taken refuge at Boulogne after being hit by a torpedo in the Channel.

Although buffeted by rough seas, the two successfully completed their task, but just as they were about to paddle away to safety the canoe was lifted on the swell and the front end became lodged in the torpedo hole of the ship. In their efforts to extract themselves, the canoe became damaged by the jagged metal and sprang a leak. With Preece paddling and Montanaro bailing, they managed to get free and the canoe limped out of the harbour. But the leak got worse, and the canoe was on the verge of sinking when the pair were picked up by the motor launch sent to collect them.

As they were hauled aboard, the limpet mines blew and an aerial reconnaissance photograph taken the next day shows that the canoeists had completed the job which the torpedo had failed to do and sent the ore carrier to the bottom. There was, however, an unfortunate ending to this latest attack: the Germans, believing the explosions to be the work of the French Resistance, rounded up 100 men and executed them. Even so, that section of the French coastline continued to remain a prime target for Commando operations, and the next one was already in planning.

This time, they were looking at Hardelot village, just south of Boulogne. The raid was handed to No. 4 Commando and had the distinction of including a party of Canadians – the first time Commonwealth soldiers had taken party in such an operation launched from Britain, and much was made of it by Mountbatten. A party of 50 Canadians were to join 100 Commandos under the overall command of Major (Lord) Lovat, who was then still second in command of No. 4.

The object was to carry out detailed reconnaissance of the beaches, capture prisoners, destroy a searchlight battery, hit a radar station and damage anything they could lay their hands on. The outcome was to be far from satisfactory, but because the eyes of the Commonwealth would be on the results of this mission – once it was known that the Canadians were involved – eventual reports of it in the immediate aftermath seemed to have not only been deliberately distorted but were, to use a modern phrase, economical with the true facts, presumably under the instruction of Mountbatten himself.

So for the real story, let us defer to the testimony of William Spearman in a long interview he taped for the sound archive of the Imperial War Museum. Spearman, a Londoner, had left the merchant navy to join No. 4 Commando just after they had returned from the Lofoten raid and so had a more-than-average knowledge of the sea and ships. The raid was staged on the night of 19 April, and according to most reports it was aborted because of bad weather. A second attempt was made on the night of 20–21 April, in which – as Charles Messenger reported in his authoritative account[2] – a new type of landing craft, the heavily armoured Landing Craft Support (LCS), was used for the first time. Bearing in mind what follows in Spearman's recollection, the use of this craft may have been linked to what happened on the first attempt on 19 April. Spearman recalled:

> Lord Lovat, in his autobiography, referred to Boulogne as a small operation on the coast of France which really didn't achieve very much but gave us a lot of experience. In my view, it was a much bigger thing than that. It was a small raid, perhaps, but I'll give you a few details that as far as I know have never appeared anywhere in print. We'd been training hard and towards the raid time we were put into Dover Castle; we were there for a week. To our amazement, there were lots of Canadians there too. As a boost to national morale, they wanted the Canadians to be involved in a raid so it turns out now that they were going to go on this raid with us.
>
> On the night of the raid, the weather was really bad but we somehow got to sea . . . in three assault landing craft tied together behind a motor torpedo boat. I never, ever remember doing this again during the war. As we approached land, German artillery picked us up and they started letting rip with tracer and shelling. At that point, they [Lovat and his staff] decided to abort the operation.
>
> We did a U-turn when they realised the Germans knew we were

[2] *The Commandos, 1940–45*, p. 135.

> coming. (And I must say here, that Dieppe was the same. We, the British Commandos, never knew whether it was Holland or France or where it was we were going, but the Canadians seemed to know in advance.) As the MTB turned round with these three boats in tow and as she got up speed to get away the ropes tightened. The middle boat was lifted out of the water and turned over. I reckon about 30 men died.

Because he'd never read the story anywhere at the time or since, Spearman began to wonder if it had happened at all. It was curiously not mentioned in Lord Lovat's account of the raid in his book, *March Past*, published in 1978. He dismissed the raid in a few sentences: 'I led a mixed force of Canadians (Fusiliers Mont Royal) and commandos on the Boulogne raid. It was an overrated affair which achieved limited success, though the Canadians did not think so . . . the Navy put them down on a sand bank and they were stuck until the tide turned. No. 4 Commando carried out all its objectives, yet I doubt if our masters were much the wiser. The raid at least proved it was possible to achieve reasonable surprise with shock troops in a defended area and some undeserved awards were handed out. But we got home in one piece, that is to say, with all weapons and wounded and gained experience.'

No mention of casualties there. But from the Imperial War Museum sound archive, another account of the same incident recorded separately by Bernard Davies, who was with Spearman on the raid, confirms Spearman's recollection:

> Sadly, we lost a landing craft, and soldiers were drowned in the rough weather. I pulled one man out of the water and I could hear his mate somewhere behind him shouting for us to come and get him, but we lost him in the darkness. We searched for an hour or so five miles off the French coast but couldn't find him. Then the E-boats came along and we lost more soldiers from their gunfire. We were sitting ducks and had to get out fast. There was nothing more we could do, and we returned to Dover.

Back to Bill Spearman's account:

> Anyway, the next night we went in again. We were all terribly surprised that we should be going back, after what had happened and especially since they [the Germans] had picked us up the first night. I suppose maybe they wanted to get Canadian troops into

action for the first time, and they seemed determined that they should go ahead with it. And so in we went again. But again, the Germans seemed to be expecting us and what they did this time – you'll not read this in any book, but it's true – was that instead of picking us up as we went in, they let us land. We got ashore on a perfect landing, which was a great surprise after the night before, but, you see, it was a trap. As soon as we landed, all this shooting started out at sea. The sky was full of tracer.

They had E-boats in the area – they are like our MTBs but faster – and they let our boats come in, and as soon as we landed all hell let loose. The E-boats made a half-circle around the beach from a distance and penned in all our boats . . . then they opened fire and our crews could not get out, not without great difficulty, anyway. There was a hell of a battle while we went in with the object of attacking our various objectives; my group had to go for a radar station, which was three miles inland, operating in small packets. As we approached the station, we reached a slit trench . . . from which a German soldier emerged and stood over us with his rifle and bayonet. He lunged at me . . . his bayonet went through my webbing equipment, which is quite thick, and through the belt and into me. I thought I'd had it; blood was pouring out. Anyhow, Joe Walsh, who was with me, shot this fellow and we carried on. I wasn't that badly hurt, although I still have the scar today.

Soon afterwards, a Very light went up and that was a strict order for us to return to the beach. We had Dutch and French money provided in case we were stranded and left behind; it was always a possibility, and we were given in advance instructions of where we would find assistance [from the French Resistance]. But it wasn't necessary. We got back to the beach and aboard the landing craft, and then out to sea and boarded the MTBs. There was still a battle going on and a shell came right through the side of our boat and exploded in the communications centre, killing two people inside. And this was the most frightening experience, having got on a ship from a raid, thinking you were safe and finding the battle still going on around you.

The Royal Navy crews managed to get the Commandos clear of the ambush and back again across the Channel without further incident. The following day, the British Commandos were surprised – and angry – to discover newspaper headlines which blared out: CANADIANS LAND IN FRANCE. They knew it wasn't entirely true. A few of them may have got ashore but, as admitted by Lord Lovat in his book, most were stuck on a sandbank. Bill Spearman and his mates were livid when they read it.

> We'd already heard they hadn't landed. The story we got was that they'd refused to because of the heavy fire, but it was only years later that we discovered the truth. Our mentions in the press came from another incident: we had this officer posted to us, a lieutenant, a very likable chap, who somehow or other lost his boots and he had to leave the ship in all that could be found – so the story went – a pair of carpet slippers. The *News of the World* made its headline: COMMANDO LANDS IN CARPET SLIPPERS. It was made into a glorified account, but in reality what use would a pair of bloody carpet slippers have been across barbed wire and so on? It was all a bit silly.

Those who worked with Mountbatten would discover that he took every opportunity to make sure of a good and positive reaction from both the newspapers and the BBC. An example was quoted to the author by Major Harry Holden-White, who was given the task of carrying out trials on a new invention dreamed up by Mountbatten's scientists: a hand-launched miniature torpedo driven by a car windscreen motor. They were used for the first time (and discovered to be utterly useless) at the Torch landings at Oran in November 1942 and, although the invasion was ultimately a success, hundreds of lives were lost as ships carrying the first wave of troops were hit by shore batteries. Harry recalled:

> When I eventually got back to England I was summoned to Mountbatten's office. I was still very bitter about the Oran business and hinted that I thought there had been a shameful waste of life. Mountbatten glared at me, but let it pass because also in the office was Colonel Robert Henriques, who handled his press relations. Mountbatten wanted me to be interviewed on the BBC and give a glowing account. I was appalled. To make such capital out of such a catastrophe seemed to me to be an act of betrayal to those who had lost their lives. There were and would be so many disasters . . . the odds were very definitely stacked against groups like us, and I wasn't going to stand there and say how wonderful it all was. Not bloody likely. Thus it was very grudgingly that he eventually signed the warrant for me to receive an MC.

Something similar can be imagined on the return of the Commandos from the first raid involving Commonwealth troops. The whole press reaction to what Lovat called a 'small raid' seemed to have been carefully managed to hide the fact that Commandos had been accidentally drowned in the first attempted landing and to bolster the role of the Canadian force. But there was another reason which went beyond placing this raid into the public domain with a reasonable hue. Mountbatten and his staff were already planning something far bigger, and in which Canadian troops were to be very much to the fore. It was to develop into one of the greatest controversies of the Second World War.

The debate rages on even in the twenty-first century, with various Internet sites devoted to recollections and criticisms, especially from the Canadians themselves. The very mention of this single word still has the potential for ire: DIEPPE.

Very soon after the Boulogne raid, it became clear that it was in fact planned as a forerunner to a major landing of thousands of troops and tanks at Dieppe. For the previous couple of months, Mountbatten had been deeply involved in a major plan known as Operation Sledgehammer, to put a striking force of several thousand men on the French coast with heavy armoury to re-engage the Germans and open up another front in western Europe to aid the Russians. Mountbatten was a key figure in drawing up proposals for the invasion on the French coast spearheaded by the Commandos, the SBS and the Paras.

In spite of pressure from the Americans for more aggressive action against the Continent of Europe, the Chiefs of Staff and the War Cabinet could not agree over Sledgehammer, either on the site for the landing or the readiness of the Allies to open up a potential battle front in western Europe. With the fall of Singapore to the Japanese and Rommel piling the pressure on the 8th Army in North Africa, the war was still not swinging towards the Allies. Eventually, Sledgehammer was postponed and then substituted with Operation Torch for landings in North Africa and put back to November 1942. In the meantime, Mountbatten was placed at the helm of Dieppe as a stopgap measure, telling the Chiefs of Staff Committee when Sledgehammer was more or less abandoned: 'It is all the more important to do one more big raid.'

Mountbatten put all other raids on hold – including Blondie Hasler's Operation Frankton. Planning for Dieppe began at the time of the Boulogne raid, and there was a certain hastiness in the air about the whole project. The chief wanted to deliver his proposals to the CoS Committee by mid-May and felt he was being frustrated by the Commanders-in-Chief of the three armed services, who each insisted they should be responsible for plans affecting the troops under their command. Mountbatten virtually pleaded for 'executive responsibility' over the next major raid; in other words, he sought total control. In spite of Churchill's support for his proposals, he was given something that fell just short of it: responsibility for 'marshalling and the launching of large-scale raids' but that his powers should end at the point of signing the operational orders.

He pressed on, and his own planners brought together their plan for the raid on Dieppe for presentation to the CoS Committee. Although it was agreed that if there were to be a raid, Dieppe was an ideal target, there were powerful detractors to what Mountbatten had in mind. The man who mattered, Churchill himself, gave it his blessing. He saw the raid on Dieppe as a dress rehearsal for the eventual invasion of Europe by the Allies, a 'reconnaissance in force', as he put it, in which landing techniques were to be put to the test in the most challenging manner.

The actual aims and objectives of the raid itself were the subject of some controversy at the time and have remained so down the years: many professed to be mystified as to what exactly was intended by it. For years it was a topic of heated conversation at military dinner tables, and an oft-quoted story is trotted out by those who could not see the point of it. It went like this: when Brian McCool, senior Military Landing Officer on the beaches of Dieppe, was captured and interrogated by the Germans, they tried for two days to get him to tell them the object of the raid. In desperation, one of the interrogators said, 'Look, we aren't fools. We know very well it was too large for a raid and too small for an invasion. So what was it?' McCool, in the manner his name might suggest, replied: 'If you can tell *me* that, I would be exceedingly grateful.'

The object was actually quite clear, at least to Mountbatten: to test all the techniques and skills acquired in earlier Combined Operations raids and apply them to a heavily manned hit-and-run assault on a major port. The troops were to be ashore for 15 hours and then make a quick exit; to all intents and purposes it was to be a model of what Churchill himself had originally called for in the creation of Commandos for raids using up to 10,000 men. If that was achieved quickly and effectively, then other attacks might follow; two, perhaps three by Christmas, thus seriously damaging the ability of the Germans to attack the vital supply lines of men and weapons from the US, now that the Americans had belatedly joined the war.

To the orthodox military mind, however, the whole scheme looked like a disaster; some mentioned Gallipoli and pointed to the grave difficulty of maintaining support for an invasion force by sea or air. The vagaries of the weather and potentially massive enemy countermeasures would undoubtedly present a challenging scenario – to say the least – and might result in the invasion force being left high and dry, taking heavy casualties.

General Montgomery himself made no secret of the fact that if there were to be a raid, he favoured an alternative proposal submitted by the planners of the Home Forces. This rival project called for a frontal attack on Dieppe, as opposed to the flank landings of paratroopers and Commandos followed by regular troops and tanks in a pincer movement proposed by Mountbatten. The latter would have avoided the high cliffs on the frontal approach and allowed troops to enter the town through the back door. In fact, Montgomery, at the first joint meeting to discuss the Dieppe raid, dismissed the Combined Operations proposals 'as the work of amateurs', and if the Allies failed to take Dieppe then the attack would for ever after be looked on as a failure. Montgomery's insistence on a frontal attack would later return to haunt those left counting the human cost of this operation.

Mountbatten argued that the occupation of the port was not the main objective, but that the first importance was to test the landing techniques and prove that a major incursion could be achieved. Montgomery remained unconvinced, and in the end Mountbatten was forced to accept the notion of a full-frontal attack and its risk of heavier casualties, although he did so on the basis that there was a preliminary air bombardment. This in itself caused consternation. There was a general observance of rules affecting the bombing of towns and cities in occupied Europe, but on this occasion the Chiefs of Staff agreed to set them aside.

There were still many misgivings about the whole project, but the success of St-Nazaire became the beacon to guide the plans for Dieppe to fruition. On 23 May Mountbatten spent three hours with Churchill and Field Marshal Alan Brooke, the new Chief of the Imperial General Staff, the latter to admit later 'we were carried away with optimism'. Two weeks later, however, while Mountbatten was away in America for talks with President Roosevelt, the Chiefs of Staff had a further meeting, at which the issue of the prior air bombardment was raised. It was now felt that heavy bombing in the narrow streets of Dieppe would block the route of tanks from the invasion force, and the RAF was also concerned that, with such a small target, many of the bombs would simply be dropped in the sea or inland. So a prior bombardment of Dieppe was cancelled.

Furthermore, Mountbatten's request for battleships to provide heavy artillery support, which would include the heavy guns of at least two cruisers, was also rejected because of the risk to important British vessels in daylight off the French coast. Naval support would thus be limited to the relatively small-calibre fire of a group of destroyers. Having lost the argument for a flank attack, then the preliminary air bombing campaign and now the presence of the navy's big guns, more setbacks followed. Mountbatten now learned from the Force Commanders, who had final say over troops, that his plan to lead the assault with heavy contingents of Commandos and Royal Marines had been reversed. Commandos would indeed go in first with the aim of knocking out German gun batteries, but the bulk of the attacking force would be made up of Canadian troops, who were originally earmarked in lesser numbers for a supporting role.

The whole of the Combined Ops planning team was aghast at this proposal, since it had always been planned on the basis of a Commando-style raid for which there had been much training and practice. Although none doubted their courage, the only 'live' experience the Canadians had of such a project was the involvement of 50 soldiers on the Boulogne raid, and most of them were left stranded on a sandbank. As Philip Ziegler pointed out, Mountbatten always maintained that this was the result of a 'high-level political decision and not one in which I was involved'.[3] One further reversal in the event would cost dear in casualties: a plan by Combined Ops to use obsolete tanks filled with explosives to blast a way through the sea wall and beach defences was rejected in the final stages of preparation by the Force Commanders. They stated they would rely on Bangalore torpedoes – long tube-like devices packed with explosives – for the job. In the event, they were to prove totally ineffective.

Mountbatten's original proposals had thus been torn to shreds by a combination of the interjection by the Chiefs of Staff and others, largely at the behest of Montgomery. Even an eleventh-hour complaint direct to Montgomery by Alan Head, one of Mountbatten's advisers at Combined Ops, received a curt reply: 'You are talking nonsense!' While many doubted the operational strength of Mountbatten's own plan, the intransigence of Montgomery and other top brass simply descended into a disturbing ego clash that undoubtedly had some bearing on the carnage that was about to follow.

[3] Philip Ziegler, *Mountbatten*, p. 189.

Chapter Seven

Dieppe: the Massacre

And so . . . after weeks of tortuous wrangling, changes of mind and the general toing and froing that seemed to surround major decisions, the Dieppe raid was now very definitely ON. In the southern ports, a massive flotilla of ships, boats and landing craft plus an assortment of hardware, which included new Churchill tanks, was assembled. The navy did its best to keep them hidden and under cover from the prying cameras of Luftwaffe reconnaissance planes, although any German agents would have had to have been blind not to have noticed the sudden increase in the military population thereabouts. A raiding force – and raid it was, not invasion – of 6,100 men was gathering. Around 5,000 were Canadians, 50 American Rangers and the rest drawn from British Commando units, including a contingent from the Royal Marines. The 2nd Canadian Infantry Division had been on the Isle of Wight since 20 May for intensive training in amphibious operations.

The raid had been set originally for the night of 4 July, but bad weather put a stop to it that night. Four days later, it was put off again, and Montgomery – soon to be heading for the Middle East – wanted it cancelled altogether because of security fears: the Canadian troops had already been briefed for the raid and with that number of men on the mission a spot of loose talk could not be avoided. Lieutenant-Colonel James Hill, who commanded the 1st Parachute Battalion, expressed his concerns. His battalion – in the original Mountbatten plan – was to parachute into Dieppe and knock out the German gun batteries to the east and west of the town prior to the major troop landings. He had gone to the Isle of Wight in May to meet the force commanders of the 2nd Canadian Infantry Division. He was clearly shocked by the talk, because it was a definite policy of the British not to brief their troops – other than senior officers – until the night of the raid, and even then they might not know their destination but would simply have got the lie of the land from models. When the 4 July movement was cancelled and the Paras were replaced by the Commandos for the assault on the gun batteries, Hill warned Lord Lovat, Commanding Officer of No. 4 Commando. Even so, and in spite of Montgomery's reservations, no changes of plan were made, except for the substitution of the Paras by the Commandos, so that the raid was not so dependent on the weather. The codename was also changed from Rutter to Jubilee.

Security was clearly an issue, and although the new date for the raid was supposedly kept quiet, the Germans were undoubtedly aware that it was imminent. In fact, they were ready and waiting. On 10 August, nine days before it happened, the commander of the German 15th Army issued an edict to his men which began with the warning that according to 'information in our hands' the Allies were coming. The order spelled out in no uncertain terms what the German soldiers could expect:

> I ask that my orders on this matter be kept constantly before them so that the idea sinks in thoroughly and they expect henceforward nothing but surprises. The troops must grasp the fact that when it happens it will be a very sticky business.
>
> Bombing and strafing from the air, shelling from the sea, commandos and assault boats, parachutists and air-landing troops, hostile civilians, sabotage and murder – all of these they will have to face with steady nerves if they are not to go under. On no account must the troops let themselves get rattled. Fear is not to be thought of. When the muck begins to fly the troops must wipe their eyes and ears, grip their weapons more firmly and fight as they have never fought before.
>
> THEM OR US
>
> must be the watchword for each man. The Führer has given the German armed forces tasks of every kind in the past and all have been carried out. The tasks which now confront us will also be carried out. My men will not prove the worst. I have looked into your eyes. I know that we are German men.
>
> YOU WILL GLADLY DO YOUR SIMPLE DUTY TO THE DEATH. DO THIS AND YOU WILL REMAIN VICTORIOUS. LONG LIVE OUR PEOPLES AND FATHERLAND. LONG LIVE THE FÜHRER, ADOLF HITLER.
>
> [signed] Your Commander,
>
> COLONEL-GENERAL HAASE

Allied troops may well have fared better if British intelligence had been similarly well informed. According to Lord Lovat, the Germans appeared to be withdrawing troops from the region but were in reality pouring more in; MI6 had failed to notice. Intelligence briefings, said Lovat, were slack and inadequate. Agents also failed to report that German field engineers had in recent times been pouring concrete into cliff excavations at Dieppe, using forced labour and prisoners of war to strengthen the defences of the precipitous terrain. They had positioned light guns and heavy machine-guns mounted on rails run from protecting cave mouths to be fired and then withdrawn to cover. Camouflaged pits were prepared at every vantage point for snipers to fire smokeless cartridges from. Land mines and barbed wire were added to the beach defences, and cliff staircases were blocked off or destroyed. The Germans did indeed know the Allies were coming.

One further mystery on the issue of security is still unexplained but is recounted by Colonel Patrick Porteous, then a captain with No. 4 Commando and who won a Victoria Cross for his bravery at Dieppe:

> It was a rather strange thing. My first wife, Louise Roome, at that time worked in naval intelligence in the Admiralty. A few days before the Dieppe raid she was going to work and bought a paper, I think it was the *Daily Mail* as far as I can remember. Glancing through, she saw an advertisement saying, 'Buy a beachcoat from Dieppe'. It was a very curious advertisement to be appearing in a national daily newspaper at the time of war, and particularly at that time. She knew the Dieppe raid was about to take place, and she took it in to her boss in the Admiralty. They looked at it and it seemed as though the shape of the dress in the picture was very like the coastline along Dieppe and there were various odd buttons and marks of various kinds which seemed to indicate the various landing places. They could never prove anything and I don't think it got any further. It's a strange story, but it might have had some significance from the point of view of the German intelligence people.

No doubt it did. As to the knowledge of the British contingents, it was non-existent, although they were well aware that something was in the offing. At No. 4 Commando, some of the men had just returned from a mountaineering course in Wales and were called back to base in Scotland in mid-July after a short leave. They were told to take the train to Weymouth, all making their own way there, and were to report to the *Prince Albert*, docked at Poole, the following day. They were sailed around the South Coast, performing practice landings on difficult beaches. Bill Spearman said Lord Lovat had scouted the area and chosen the most steep and rocky places:

One of them was close to St Ives, Cornwall, and was a really lively spot, very rocky, steep and with the waves crashing ashore. The reason, we discovered, was that our practice landings were to be witnessed by top brass – we were going to give a demonstration. So Lovat picked the roughest bit of coast where on a certain type of day – and certainly on the day we did it – the tide fall was 12, 14 feet. We didn't have landing craft that day but were using flat-bottomed ferry boats with about 30 men aboard each one; the white spray was all around us as the waves came crashing into the rocks, and the boat reared up like a stallion and momentarily is left hanging in midair as the water drops away, then it comes down with a frightening bang. The object was at that moment, when it was high, to jump out on to the rocks and climb the cliff. Well, in front of the generals and some Americans, it turned into a fiasco. One of the boats overturned and two men were drowned. Even so, they made a training film of it – called, appropriately enough, *Rough Landing*.

It was now pretty obvious that they were soon to be on the move, although Mancunian George Cook, then 23, remembers that still nothing had been said about their eventual destination. They just carried on with their practice landings and other training routines:

After about ten days or so on board the *Prince Albert*, we came off the ship and went into digs in Weymouth, where we resumed normal training. Then we were told one night when we'd finished: 'Settle your landladies, pay your bills and be on parade at six o'clock in the morning.'

When we went on parade there was a line of lorries, covered-in army trucks, which we climbed aboard and set off to a camp outside Southampton, where we had a meal. Still nothing had been said as to where we were going. We then got back in the trucks and went into Southampton docks, where we rejoined the *Prince Albert*. Lord Louis Mountbatten came aboard and then it was announced we were going to the French coast, what we had to do, what each troop had to do. We were told if anybody got wounded to leave them where they fell, not to bother with anybody. The medics would deal with them. We had to press on to our target, a gun battery which had to be destroyed before the Canadians could land. Anyhow, Mountbatten spoke to us en masse. He said he wished he were coming with us. It was a very nice talk, and we cheered him, then we started drawing ammunition. Our troop was doing the demolitions, so we drew explosives and we'd a fair amount of stuff which we packed up, ammunition, spare Bren pouches, grenades which we'd all primed, and then another meal and we set off, sailing down the Solent and past the Isle of Wight. It was a beautiful evening for it . . .

The Jubilee plan called for attacks at five different points on a front of ten miles along the beaches of Dieppe, trimmed as they were with white chalk cliffs. There would be flank attacks to the east and west of Dieppe and a full-frontal attack towards the main town itself. The four simultaneous flank attacks were to go in just before dawn, followed half an hour later by the attack on Dieppe supported by the 14th Canadian Tank Battalion. Canadian troops would form the force for the main attack as well as landings at the cliffs at Pourville two and a half miles to the west and at Puys to the east. Two coastal batteries codenamed Rommel and Hindenburg were sited to the east and west of the town, and these were to be hit by the Royal Regiment of Canada supported by the contingent of 370 British Royal Marine Commandos.

The role of the British Nos. 3 and 4 Commandos was principally a quick smash-and-grab raid – as Lovat termed it – to get in ahead of the main group and destroy the batteries at Berneval on the eastern flank (No. 3 Commando's target under Lieutenant-Colonel John Durnford-Slater) and at Varengeville in the west (No. 4 Commando under Lord Lovat). Without those guns being silenced, it was judged that the main force would not be able to make their entrance. In fact, the batteries were only a part of the massive German defences on hand. Their extent was so seriously underestimated by intelligence sources and aerial photograph reconnaissance that even with the batteries knocked out, a massacre was waiting to happen.

The first part of the plan went awry even before the men were able to land. After sailing across the Channel in an assortment of ships pressed into service for the mission, they would transfer to landing craft five miles offshore for the final approach. As Patrick Porteous has said:

Our unit sailed in an old Belgian cross-Channel boat which had been fitted with landing craft (assault). There were eight craft on the ship, four on each side. We set off just before dark with an enormous convoy. There was a great mass: hundreds of ships and landing craft as far as we could see. Minesweepers went ahead to clear a gap through the German minefield which was about ten miles offshore.

We got to lowering position (for the landing craft) at about 4 a.m., having had a good night's sleep and a nice breakfast of stew. All grenades had been primed, magazines filled; everything was all ready. We were lowered away under the protection of a steam gunboat[1] commanded by Lieutenant-Commander Peter Scott, the naturalist. He had a flotilla of these gunboats, only one of which we actually came in, and there was also an armed motor launch as well, which came in with us. In fact, these were our only local support going in. It was about a two-hour journey to the landing point. Luckily, the sea was very calm, very clear. After we'd been going for about half an hour or so we suddenly saw some fireworks going on way out in the east on our left flank.

This, it transpired, was No. 3 Commando, who had been going for the other battery at Berneval and bumped into a German naval convoy and flak boat coming down. There was quite a firefight. The landing craft of No. 3 Commando were all scattered, and the whole thing was a bit of a shambles there.

A shambles indeed. No. 3 Commando, which had been billeted at Seaford and had had no contact with No. 4 before the off, were moving towards their target zone in 24 R-boats, or LCPs (Landing Craft Personnel), and supported by LCFs (Landing Craft Flak), which carried two four-inch guns. Durnford-Slater and his headquarters team were travelling in the steam gunboat, which led this mini-flotilla for protection. All was going according to plan until a German flak ship, leading a small convoy of other enemy craft, spotted the No. 3 flotilla and opened up with tracer. The steam gunboat was badly mauled and was quickly put out of action. At least half the men on board were killed or wounded, and the ship limped away and was later taken home under tow.

In the onslaught, the landing craft carrying the bulk of the No. 3 personnel were scattered and many boats were sunk and their occupants killed or drowned. Only four of the twenty-four landing craft made it to the shore, and they came under heavy fire as they landed. Many more troops were killed or wounded at this point, and only 18 men remained to fight their way towards the battery at Berneval. They were ridiculously outnumbered by the German positions ahead of them and were forced back to the beach after battling it out for more than two hours. There they discovered that the tide had gone out and their remaining landing craft were high and dry. Only a handful managed to escape, the rest being taken prisoner.

[1] The old steam gunboats, larger than a motor gunboat, were hardly conducive to any surprise landing: they had an unfortunate habit of emitting an explosion of sparks and smoke from the funnel, which was generally followed by a tirade of abuse down the speaker tube from the bridge to the engine room.

No. 4 Commando had better luck. Lovat divided his party into two groups, one to lead a frontal attack on the battery and the other to go in from behind. Peter Scott took his steam gunboat in as close as he could, firing as he went. The landing went successfully, but two very steep narrow gullies leading to the battery were blocked by a mass of barbed wire. One was impenetrable, but the other was blown apart with a Bangalore torpedo and then a path overlaid with rolls of wire netting that the men had brought with them for the purpose. The German troops in defensive positions around the battery were now aware there was a landing on their section of the beach and heavy fire rained down on No. 4 from a gun pit up above which was heavily camouflaged. A two-inch mortar was fired and, said Patrick Porteous, 'by some incredible fluke they managed to land one in the gun pit, where a pile of cordite charges for the gun ignited with the most tremendous explosion and flash, and the guns didn't fire again after that. The gun pit was all on fire; they had camouflage nets and so on over the top in an awful mess. All the crew of that particular gun were killed.'

Guns were firing at this stage at the ships bearing the main attacking force now heading towards the central area. A German observation post was on a nearby lighthouse, which was flashing, and two men were dispatched from No. 4 to cut the telephone wires from the battery to the lighthouse, shinning up a telegraph pole. Meanwhile, the two parties from No. 4 began to make headway towards their target, as George Cook described:

> The lighthouse was still flashing, so I thought, we're going to have a pretty bad welcome here. They were firing tracer from some pillboxes and Lord Lovat, as he went over the wire, turned to me and said quite casually: 'They're firing too high.' Lord Lovat was about six foot; I'm five foot four, so I thought, if they're firing over his head there's no danger they're going to touch me. In fact, they did fire some mortars, and four or five of our blokes went down before we got off the beach; some were wounded, some killed. One of the medical orderlies, Jimmy Pasquale, stayed on the beach with these chaps who were either dead or dying to do what he could for them, but we pressed on inland. As we were going in, there was the skull and crossbones and it said, 'Achtung Minen!' The beach was mined but we just ran straight on. I didn't hear any mines go up so obviously nobody was getting blown up. The firing continued as we came off the beach. I was up in front of the group.
>
> In the middle of all this there was a little tiny bungalow, and incredibly a woman came rushing out saying, 'Bonjour', waving a

bottle of wine to give us a drink. But we'd no time and just saluted her and ran on. Then we all went our diverse ways. I was with a group which consisted of Sergeant Desmond, Sergeant Home, Lance-Corporal Diblock, Private Tommy Branwell, myself and Captain Porteous. We came to a house which we thought might be a refuge for Germans. Tommy Branwell started to shoot the lock off when I noticed an air-raid shelter and some people looking out. It was the owner of the house, and they came out and he told us: 'No Germans here; they're over there.' So we told him to get back in his shelter and stay there otherwise he'd get shot. We went on and then for the first time I spied the battery. I actually heard some shots being fired. I could hear the boom, boom, boom. Then I saw a chap come out of a hut all dressed in white. It turned out he was a cook. He did a piddle at the end of the hut. He stood there urinating and I thought, blimey.

Then suddenly there was some firing, and one of our parties started firing on to the battery. I think they'd been spotted, so the Germans were having a go. Somebody, I think it was Spearman, shot a bloke out of the ack-ack tower. He did a lovely swallow dive 60 feet off the top. There was some firing coming our way as we arrived at the battery in an orchard, and Sergeant Home and myself reached the barbed wire and began to cut it. I heard a faint 'Ugh' and when I looked up there was Sergeant Home staggering with blood spurting out of his chest. I had a look at him and he was dead, which was a bit of a shock to me because he was about the toughest fellow I ever knew. He'd been on the North-West Frontier and in Palestine and he knew his stuff. I thought, I'll stick by him. Anyway, I cut the wire and I went back to Porteous and reported: 'Sergeant Home's dead. He's been shot.'

The next thing I knew . . . bonk, that was me. I got hit in the face, in the shoulder, must have been a machine-gunner, I think. I went down, poleaxed, and I knew nothing then. When I woke up, came round, the sun was shining on me. I'd lots and lots of flies. I tried to move and couldn't, tried to get up but I'd lost that much blood I was soaked. Just as I awoke and tried to gather my senses, three Germans arrived. One of them stuck his bayonet right on my throat and one said to me in English: 'Can you move?' I said: 'No, I've tried. I can't.' They then went through my pockets, took my cigarettes and promptly lit up. The one who spoke English asked me if I wanted anything. I said I wanted a drink. There were two Frenchwomen who had come on the scene, so he told them to get me a drink. He said: 'We'll be back for you.' So he left me. One of them was an older woman. She cradled my head in her lap; the other one came out with something in a cup and tried to give me a drink. Part of my face and lower jaw had gone, just a big hole there. I couldn't drink, so they came back with a glass and it had wine in it. They spooned it down into my throat. Then they left me. The next thing I knew after that was in the cattle truck along with a lot more. I must have passed out after I'd had the drink, and it was dark and we were in this cattle truck; other chaps were moaning, Canadians mostly . . . then I passed out again.'[2]

The raiders of No. 4 Commando carried on. They knew their orders: all those wounded or killed were to be left where they fell. Although suffering a number of casualties, the Commando had managed to put a large number of men into the final assault from their original contingent. Patrick Porteous was seconded from another troop not taking part in the attack but eventually found himself commanding F Troop, which was charged with the assault on the rear of the battery:

> We went up the bank alongside a river. We were wading most of the way because the Germans had flooded the valley there. It was fairly heavy going for about a mile directly inland. Then we swung left across open ground behind the German defences, which brought us round to a little wood at the back of the battery we were to assault. There were two troops formed up who were going to do the assault from there, B Troop and F Troop. I was with F Troop, and we were on the left going in. We managed to get in there without any problem at all. As we got into this area of the battery, we bumped into a truck of German soldiers who were just disembarking. Obviously, they'd come up from Ste-Marguerite, probably, or somewhere down in that direction. We managed to knock them off before they got out of the truck, killed the lot of them, virtually, with Tommy-guns. We then started working our way through this very dense bit of country, all these little cottages and hedges and so on. Roger Pettiward, who was the troop commander, was killed by a sniper, and the other subaltern, John Macdonald, the other section

[2] George Cook was eventually transported to a German military hospital in Paris. He was in a coma for a month and eventually underwent extensive reconstruction surgery on his face, with bone and skin grafts conducted by skilled German surgeons who were also treating their own seriously wounded men. He remained in hospital for almost a year before he was taken to a prison camp in Germany which had been set aside for those captured at Dieppe. He was repatriated towards the end of 1943, full of praise for the medical treatment he'd received.

> commander, was killed by a stick grenade hurled at him, which left me in command of the troop. I was going along a little lane towards the battery with a bank on the left and I suddenly saw a German popping up on the other side of the bank. I threw a Mills grenade at him and he threw a stick grenade at me. As soon as they went off, I popped up to take a look but unfortunately he popped up a little bit quicker and he shot me through my left hand, so I withdrew and put on a field dressing. We pressed on and there were several other casualties, chaps who'd got sniped or blown up with grenades. We lined ourselves along a bank, which was giving us a little bit of cover from where the actual guns were. I made contact with Commando headquarters, who were approaching the battery area, and he said: 'Come on, it's time we went in.'

F Troop with B Troop charged forward with bayonets out front into the gun pits. The 90 metres or so of ground to be covered seemed like a mile. Patrick Porteous took a second bullet on the way, this time through his thigh, which slowed him down as blood began pumping from the wound. They got into the gun pits, the first of which had had the mortar bomb in it and was full of corpses. Porteous staggered on to the next gun pit and then collapsed through loss of blood. Meanwhile, the final assault on the battery continued, and Donald Gilchrist has described it in colourful prose:

> Our section doubled forward. Close to us a hedge rustled. The Bren leaped in Marshall's hands. I glanced sideways in surprise at a grey form settled in a huddled heap. Up went the grenades to explode in the right-hand gun pit. With fixed bayonets, F Troop attacked, yelling like banshees . . . Razor-sharp Sheffield steel tore the guts out of the Varengeville battery. Screams, smoke, the smell of burning cordite. Mad moments soon over. And a rifleshot from the buildings behind the hedge: lying in the yard was a wounded Commando soldier. From the gloom of a barn emerged the German who had cut him down. He jumped up and crashed his boots on the prone face. Our weapons came up. A corporal raised his hand. We held our fire. The corporal took aim and squeezed the trigger. The German clutched the pit of his stomach as if trying to claw the bullet out. Four pairs of eyes in faces blackened for action stared at his suffering. They were eyes of stone. No gloating, no pity for an enemy who knew no code and had no compassion . . .

A demolitions team with made-up charges called plastic 808 moved in. They opened the breech of the guns, shoved in the charges and closed the breech, having set the delay mechanism for two minutes. So all the guns were loaded with a 100-pound shell that exploded with the plastic and opened up the guns like a banana skin, so they were completely out of action. Having blown the guns, No. 4 Commando had completed its task and now had to get the hell out of the place. According to Bernard Davies, a veteran of Dunkirk who joined No. 4 on his return, they left behind about 400 Germans, dead or wounded. They took a number of prisoners who were used to carry Porteous back to the beach on a makeshift stretcher ripped from an old shed. The prisoners didn't like it because mines had been laid there and they weren't quite sure where they were, so they were very nervous. Porteous and No. 4 were homeward bound, leaving behind 45 of their number, dead or missing, out of its original force of 265 men. No. 3 Commando suffered far worse losses. They had been cut to ribbons: 117 killed, wounded or missing out of 420.

At the beach, the returning Commandos of No. 4 could see the massive concourse of Allied shipping far off the coast and the landings at Dieppe itself were already underway. Porteous remembered:

> An assault craft laid a smoke screen for us to get away. Our landing craft were standing well off so that they wouldn't get grounded on the beach with the falling tide. We had to wade out almost shoulder-deep. I was loaded by these two poor Germans who were scared stiff; they thought they were going to have their throats cut or something. I was then transferred first to Peter Scott's steam gunboat, where there was a medical orderly who tied up my leg and hand as best he could, and then transferred to a destroyer. There were six hunt-class destroyers covering the whole operation. That was the only naval support we had – just these little 4.5 guns which were not very effective, frankly. The destroyer I was on went on trying to pick up wounded from the awful mess that was going on outside Dieppe; it was an absolutely appalling shambles. When we were fully loaded with wounded, we had one bomb very near us from a German aircraft. All the lights went out and it was all rather frightening. We eventually returned to Portsmouth at about one o'clock the following morning, having started off about 24 hours before. I was carted off to hospital for six weeks and later discovered I was to receive the Victoria Cross. I could never understand why. I was very lucky, I think, extremely lucky to get the award.

The modesty of Porteous, who was immediately promoted to major, was apparent from the citation for his VC, which recognised the fact that he had led the unit which achieved the most successful part of an otherwise catastrophic operation. It read:

> Major Porteous was detailed to act as Liaison Officer between the two detachments whose task it was to assault the heavy coast defence guns. In the initial assault Major Porteous, working with the smaller of the two detachments, was shot at close range through the hand, the bullet passing through his palm and entering his upper arm. Undaunted, Major Porteous closed with his assailant, succeeded in disarming him and killed him with his own bayonet, thereby saving the life of a British Sergeant on whom the German had turned his aim. In the meantime the larger detachment was held up, and the officer leading this detachment killed and the Troop Sergeant Major fell seriously wounded. Almost immediately afterwards the only other officer of the detachment was also killed. Major Porteous, without hesitation and in the face of a withering fire, dashed across the open ground to take over command of this detachment. Rallying them, he led them in a charge which carried the German position at the point of the bayonet, and was severely wounded for the second time. Though shot through the thigh he continued to the final objective where he eventually collapsed from the loss of blood after the last of the guns had been destroyed. Major Porteous' most gallant conduct, his brilliant leadership and tenacious devotion to duty which was supplementary to the role originally assigned to him, was an inspiration to the whole detachment.

Elsewhere, there was bravery aplenty too, but the outcome was already heading for the history books: a massacre no less was occurring on the beaches in front of Dieppe under a barrage of fire from German artillery and troops while overhead a massive air battle raged between Allied planes and Luftwaffe, the latter coming in wave after wave with bombers and fighters to attack the Canadian troops now coming ashore and their boats and ships. What followed was 'ten hours of unadulterated hell'.

Those troops that managed to get ashore were swept by incessant fire from well-protected enemy positions on the adjacent cliffs. For the troops, disaster overtook disaster as the attacks began first on the two flanks and then with the full frontal on to the beaches of Dieppe itself. The Royal Regiment of Canada, which was to hit the Rommel battery, formed up behind the wrong gunboat and with their start delayed by 20 minutes they were caught by searchlights as they landed. With the element of surprise long gone, they were literally cut to pieces. Only 60 of the force of 543 would return to England that day. Three platoons of reinforcements from the Black Watch (Royal Highland Regiment) of Canada were pinned on the beach by mortar and machine-gun fire, and were later forced to surrender. Of those who landed, 200 were killed and 20 died later of their wounds; the rest were taken prisoner in what was the heaviest toll suffered by a Canadian battalion in a single day throughout the entire war.

At Pourville, the Canadians, the South Saskatchewan Regiment and Queen's Own Cameron Highlanders of Canada, fared better at first, then German opposition mustered quickly as they crossed the River Scie and pushed towards Dieppe. Both units were halted well short of their objectives by intense German fire and overhead strafing from the Luftwaffe, and in fact the losses in withdrawal were heavier than in the advance. Ammunition was running out, and many were forced simply to surrender.

The main attack was to be made across the pebble beach in front of Dieppe and timed to take place a half-hour later than on the flanks. German soldiers well dug into clifftop positions and sniper spots in the hotels along the esplanade were waiting. As the men of the Essex Scottish Regiment assaulted the open eastern section, the enemy swept the beach with machine-gun fire. If someone, somewhere had at that moment shouted 'Remember Gallipoli' they would have been absolutely right; but no one higher up the chain of command had remembered, or if they had it had been carefully placed at the back of their minds. In many respects, Gallipoli was being played out again and again. The similarities were remarkable: then it was Anzac troops and Gurkhas, this time it was Canadians and Commandos. The British navy assault on Gallipoli was withdrawn because the flotilla commanders were afraid of losing their ships; at Dieppe no big guns were provided at all, and only the small guns of the destroyers were on hand to provide pretty ineffectual support.

The assault on Dieppe depended entirely on the success of the four flank parties, just as the Gallipoli attack relied on flanking attacks. If they failed – as three of them did at Dieppe – the beach was an absolute killing ground. The headlands on either side provided the Germans with open range targets, firing straight down and enfilading down the beach. The incoming troops never stood a chance.

All attempts to breach the sea wall were thrown back with serious losses. The reserve battalion, Les Fusiliers Mont Royal, was sent in and suffered the same fate. The Royal Hamilton Light Infantry landed at the west end of the promenade opposite a large casino and forced their way through into the town, where they met vicious street fighting. The new Churchill tanks, a whole regiment of them, foundered and spluttered.

Those that did not sink in the sand lost their tracks on the huge pebbles that littered the beaches. Only twenty-four managed to get fully mobile, and only six successfully negotiated the sea wall, but they, too, were soon left smouldering. Nevertheless, the immobilised tanks continued to fight, supporting the infantry and contributing greatly to the withdrawal of many of them; the tank crews became prisoners or died in battle.

The last of the Commando troops were still to go in. They were from the No. 40 Royal Marines Commando, most of whose 370 men, mustered for what was their first major action of the war under the Commando banner, had been kept offshore in reserve and who were now to be confronted by the fate of the Canadians. They began their move forward at about 1130 under cover of a smoke shield up to about 450 metres from the shore.

As they emerged from the smoke, the Germans opened up with shellfire at about 365 metres and then light machine-gun fire. The first of their landing craft broke down at that point, floating aimlessly amid the shells landing around it. The navy crew managed to start the engine again, and the craft limped on for a few more metres before it finally gave up the ghost with 275 metres still to go. Led by their commanding officer, Lieutenant-Colonel Picton-Phillips, the party was by then nearing the beach under heavy attack. Seeing the disaster that was about to confront his men and the sheer suicidal weight of German firepower, he put on a pair of white gloves, stood up in the boat using semaphore and waved them to return to the protection of the smoke. He was shot and killed within seconds and his own craft, hit broadside on, burst into flames. His second in command, Major Houghton, was wounded and taken prisoner. In spite of the warning, the marines continued to land. One of them, Jim Hefferson, recalled the scene: 'There were bodies everywhere, with a yard-wide river of blood on the edge of the sea. Craft and tanks were burning, planes were diving overhead and mortar fire was raining on the beach throwing large stones everywhere off the shingle.'

Three of the marines' craft actually managed to beach, and all those abroad were either killed, wounded or captured. They included their second in command, Major Titch Houghton, who was killed along with several of his men as their assault craft was blown to pieces, hit by mortar fire as it landed. Not one of the men in those three craft would return home to base that day, and the toll would be heavy: 76 out of a total force of 370, even though many did not even get ashore.

The remaining Commando forces were made up of No. 10 (IA) Commando, which included a French detachment, and X Troop, made up of former German nationals, the latter captured to a man and never seen again; they doubtless died a horrible death. No. 10 Commando lost six out of the eighteen members landed, as did the US Rangers who also lost six out of eighteen. One other Commando formation also made its first – and equally disastrous – appearance at Dieppe: the Royal Navy Commandos (C and D), which operated as beach masters, with a party assigned to each of the beaches. In fact, in common with the rest of the debacle, some could not reach their assigned beach and suffered heavy casualties en route.[3]

Operation Jubilee was all but over by lunchtime. Those who could make it back to the boats left behind a battlefield the like of which none of them ever wanted to see again, the long beachfront of Dieppe littered with bodies and smouldering wreckage. Of the 4,963 Canadians who embarked on the operation, only 2,210 returned to England, many of them wounded. When one considers that large numbers of the Canadian troops never actually got ashore, the percentages of death and survival were totally unacceptable in any war. There were in all 3,367 Canadian casualties, including 1,946 prisoners of war and 907 Canadians dead.

The cost was, of course, far greater when the toll from all the other groups involved – fielding around 3,000 support personnel – was added in. The Royal Navy crews manning the ships, motor launches, boats and landing craft suffered heavily, as did the Allies' air force. The RAF supplied 74 squadrons and the Royal Canadian Air Force 12. The extent of the air battle over Dieppe was evident from the losses suffered by the Allies: 106 planes shot down, with only 30 pilots surviving, against Luftwaffe losses of 96 aircraft. Other hardware lost by the Allies included 300 landing craft and launches, 28 tanks destroyed or abandoned and 1 destroyer gutted.

One of the Allied pilots came down in the sea after a ten-minute dogfight with a German fighter. His luck was in that day. The pilot, a Canadian, was spotted drifting seawards by the crew of one of the landing craft taking No. 4 Commando away to safety. They changed course to pick him up, hauled him into the boat, someone stuck a tin of self-heating soup in his hand and a voice said: 'How's that for service?' The pilot was lost for words, and the casualty toll was cut by one.

The recriminations began on the homeward journey, but the PR machine in London was already spinning a tale of success. Mountbatten's communiqué spoke of great lessons learned for the future, meaning the eventual invasion of mainland Europe, and accentuated the positive

[3] Those captured included a group who went on to become famous as the real-life players in the famous story of 'Albert RN', using a dummy on parade to fool German prison camp officers while they escaped. By the end of 1943, 22 RN Commandos had been formed to provide beach cover for virtually every major Allied landing in the remainder of the war.

by pointing out that two-thirds of the attacking force had returned home. Lessons were undoubtedly learned – or relearned, because they had been learned before, at Gallipoli – but whether it was worth the cost was quite another matter, about which opinions were deeply divided. The debate would go on for years, with particular bitterness aroused in Canada, quite naturally, whenever the topic of Dieppe arose. None of the leading personalities involved would ever admit to failure or defeat, but each of them must have felt some guilt, even shame, for what had happened. Lying in his hospital bed for six weeks in the immediate aftermath, Patrick Porteous had time for reflection and came to the conclusion shared by many: Dieppe was an absolute disaster that should never have happened. As he read the newspaper stories, and later the autobiographies and biographies of wartime military leaders, one of the oft-repeated statements that annoyed him most was the very one that led the Mountbatten post-Dieppe statement about lessons learned. 'My feeling was,' said Porteous, 'that 90 per cent of those lessons could have been learned training in Britain on the beach at Weymouth or anywhere. The Churchill tanks, for example, which had never been in action before and just shed their tracks on the shingle: they could have had endless trials in Britain, to see whether they could work over shingle. I'm very, very bitter about it.'

Who was to blame? Years later, in his memoirs, Field Marshal Montgomery stated that the two biggest mistakes in the raid were the decisions to substitute the paratroops with the Commandos and the cancellation of the preliminary bombing raids on German defences. Mountbatten was furious and arranged for Wildman Lushington, his former Chief of Staff, to write a letter to him pointing out that Montgomery himself was chairman of the meeting at which those two decisions had been taken. He then copied the letter to Montgomery, enclosing it with his congratulations on his 'magnificent contribution to the highest level of military history'. Montgomery, his arrogance undiminished, replied that he could not remember Lushington – 'he is obviously a little man' whose views he presumably regarded as being of no consequence.

The war of words continued with their respective biographers. Nigel Hamilton in his book *Monty: The Making of a General*, published in 1981, described Mountbatten as 'a master of intrigue, jealousy and ineptitude, like a spoilt child who toyed with men's lives with an indifference to casualties that can only be explained by his insatiable, even psychopathic ambition'. He went on to claim that the real reason for the debacle was that Mountbatten had promised naval artillery support at Dieppe which was never provided, other than the limited capabilities of the hunt-class destroyers. Philip Ziegler, Mountbatten's biographer, responded, stating that Mountbatten was well aware of the shortfall in naval firepower and had plagued naval command for battleships or at least cruisers to back up the raid. This was refused.

Ziegler meanwhile suggested that Hamilton might have been better to have examined the faulty intelligence surrounding the Dieppe preparations – a fact confirmed by officers returning from the raid. Even the official battle summary placed first and foremost at the top of its list of failures the 'absolutely mistaken' assessment of German defences.

The debate resolved nothing. Accusations and counteraccusations merely heightened the lingering bitterness. There was, however, another level on which this catastrophe was played out, the secret one in the mind of Churchill himself. He in fact performed something of a miraculous deception by allowing the raid to go ahead, knowing full well that there were weaknesses. He could have stopped it. Montgomery himself wanted it stopped. But Churchill was under great pressure from the Americans and Russians to open a second front in western Europe, to relieve the German onslaught against the Soviets. He knew that the Allies were not ready for such a move and was certain that any landings in Europe would be beaten back. Churchill was not going to be browbeaten by Roosevelt or Stalin into another, and far more costly, humiliation like Dunkirk, especially as Montgomery was about to launch the 8th Army into a new and vital offensive against Rommel's Afrika Korps (as is often overlooked). The risk, as Churchill recalled in volume IV of *The Second World War*, was that with Stalin facing annihilation on all fronts, he might have sought a separate deal with the Germans, as Lenin had done in 1917. The threat was implied when Stalin sent his Foreign Secretary, Molotov, to London and Washington in that summer of 1942 to plead for a second front to 'draw off 40 German divisions'.

Churchill was having none of it. While the generals, Chiefs of Staff and the War Office bickered among themselves, he let the whole thing take its course and was taking a swim in the Mediterranean during a visit to the Middle East at the time of the raid. Dieppe was a show of action at a difficult time. It could have been better if the top brass had spent a little less time polishing their egos. However, it quietened the demands for a second front and indeed forced the Germans to bolster their defences in western France. What Dieppe proved, more crucially, was that any Allied landing against the ports of western Europe was for the time being an impossible dream. The key lesson that was learned would only come into play on the D-Day landings: that an invasion had to be achieved on the widest possible front and not, first and foremost, against the most heavily defended positions. That discovery, as Churchill argued often, would save tens of thousands of lives in the future. 'Strategically,' he wrote, 'the raid served to make the Germans more conscious of the danger along the whole coast of occupied France. This helped to hold troops and resources in the west, which did something to take the weight off Russia. Honour to the brave who fell: their sacrifice was not in vain.'

Chapter Eight

Hitler's Revenge

For those captured at Dieppe, there was to be a continuing ordeal that went even beyond the usual unpleasantness and deprivation of the German PoW camps. It began as a result of the first of a series of incidents that led ultimately to the infamous edict issued by Hitler as set out in the prologue to these chapters. Patrick Porteous believed that it was started when among those captured was a Canadian brigadier who was found to be in possession of a copy of the full operational orders for the D-Day invasion. Within those orders was a reference to German prisoners taken by the Allies on this operation: they were all to have their wrists fettered. Exactly why this was decided, or by whom, is unclear.

Soon after the Dieppe raid, it was alleged in the Berlin newspapers that a number of German soldiers found dead on the beaches at Dieppe had their hands tied behind their backs. It was further claimed that some had been shot in the head. The story got around that this was the result of cold-blooded executions by Lovat's No. 4 Commando whose men, it was said, had been instructed not to take prisoners. Patrick Porteous strongly refuted the suggestion: 'There were no prisoners taken in the main part of the Dieppe raid as far as I know; they were too busy saving themselves. We only had three or four prisoners on our part of the operation, and they came with us, carried me in fact, and certainly did not have their hands tied. But it [the document] was of course wonderful ammunition for Hitler, who then kept all the prisoners from Dieppe in shackles for the next six months or so.'

The immediate upshot was that Hitler personally ordered that 2,500 prisoners of war from Dieppe, including the Canadian troops, British Commandos and other personnel captured during the raid, should be quarantined in specific camps and chained together. In Britain, the War

Office immediately issued a statement denying that any German prisoners were fettered or shot outside the heat of battle. However, in a tit-for-tat move, some German prisoners were shackled as a reprisal by both Britain and the Canadians. Winston Churchill himself said of the issue: 'In his fear and spite Hitler turns upon the prisoners of war who are in his camps and in his power. To show weakness of any kind to such a man is only to encourage further atrocities.'

The denials, however, were to shield the truth: British Commandos did indeed shoot prisoners whose hands had been tied behind their backs. Captain Porteous, who had been severely wounded, it will be recalled, and was being carried under fire by German prisoners to the landing craft that would take him home, may not have been fully aware of exactly what was happening around him. That, chapter and verse, unfolds now from the testimony almost half a century later of William Spearman[1] who, it must be said as a prefix, was one of Lord Lovat's most ardent fans:

> When we hit the gun batteries, there was total confusion, really. It was pitch dark, gunfire everywhere and planes overhead, and anybody who was not a Commando you recognised, you just killed him. We all had instructions to bring back prisoners, but in the heat of battle on the kind of operations that we undertook it really was a case of kill or be killed. I mean, you don't stand a chance of taking a prisoner. We knew there was a garrison of 300 or more Germans right there, manning those guns, and we had to get in there and put a stop to it at all costs. You can't have a debate with yourself – 'Now do I shoot you or try to take you prisoner?' – when mortar fire is whizzing past your ears. Nor is there an opportunity to stand there with a megaphone and shout: 'Look, we've got you surrounded. Would you like to surrender?' You just go in and do your job and hope you don't get shot yourself. A lot of people did, unfortunately. We had to put them out of action and there was no stopping us.
>
> Anyway, at the end of it we got through. When all the resistance had finished, and a certain amount of people had prisoners, Lovat called me over with another fellow, Percy Tombs, and asked us to find the safest route out because we had prisoners to take with us and we had wounded to carry. We'd be slowed down by walking wounded and the stretcher wounded; people were carried on made

[1] William Spearman, Imperial War Museum Sound Archive, Accession No. 009797/08, Reel 4.

up bits of wood. Percy Tombs and I found the best route, and we all made our way back on to the beach to look for the boats. When we got down . . . you saw this bloody long beach. Maybe it was because we were tired, but it seemed a hell of a long way to go when you're under fire. The Germans were on top of the cliff by this time, and as soon as you got on the beach they started shooting you up. There was also a problem with the boats coming in under fire. I remember seeing the splashes in the water, but we had to get away from the open beach otherwise we would be cut to ribbons.

But there came a point when it was realised there were too many people to get in the available boats. And because there were too many people, a decision had to be made as to what to do with the prisoners. I wasn't, of course, involved in any of this decision-making. I was just a bod there on the beach. But what happened in the event was that they decided the only thing they could do was to shoot some of the prisoners.

Were their hands tied behind their backs?

Yes. I can't say exactly how that came about. But if we were to lumber ourselves with prisoners from the battery and get them back to England, we had to put them under some sort of control where they wouldn't be a burden to us or hold us up. So our plan was that we would put their hands behind their backs and at the base of the thumbs tie them tight with very fine twine, so that if they were to make any effort at all it would almost sever their thumbs. So this is what we did in order to get them back. We couldn't have got them back in any other way. They just moved along then under their own steam with the rest of us.

Were they cooperative in going back or were they obstreperous?

If they'd been in any way uncooperative, they wouldn't have lived to know it. There was no way they could be uncooperative. They were tied up; it was all done in a minute and they were all taken back. But because we were limited with the boats, there were too many and they had to be shot. And they were shot and they were left on the beach. This was unfortunate because there was no way that any of us would have enjoyed shooting prisoners. I mean, it's not in the nature of an Englishman to do that. Even though the Germans and the Japanese did some terrible things, I don't think that

> anybody would enjoy shooting a prisoner, especially people like us, because we weren't fighting because we were killers, we were just fighting to complete an operation.

Hitler's own investigations actually proved what Spearman was able to confirm so long after the event, and the dictator's fury was exacerbated over the coming weeks when more raids were planned and carried out in a particular manner that seemed to have the capacity to needle the Führer.

No more major raids on the scale of Dieppe were contemplated. The main focus now was on an invasion – Operation Torch, which would bring the Allied landings in North Africa in November 1942. Meanwhile, Mountbatten's Combined Operations planners switched their attentions back to small-party raids and a number of groups based on the Commando ethos continued to proliferate, with schemes to attack enemy-held territory around the Mediterranean, the French coast and the Channel Islands.

Among them was the Small Scale Raiding Force, which was given a cover name of No. 62 Commando. The force was actually based on one formed in the autumn of 1941 from B Troop of No. 7 Commando. Two of its brilliant and inventive young officers, Captain Gus March-Phillips and Lieutenant Geoffrey Appleyard, were seconded to train newly recruited agents in the Special Operations Executive in the various arts of spying, sabotage and assassination, for which specialist skills at that time had been acquired only by Commandos. They also needed a boat to land the trained agents at their various destinations, a role incidentally also regularly carried out by the SBS. It was not unusual around the Med to see a British submarine rise to the surface and on to the casing would step a man in a suit carrying a briefcase who would then be delivered to shore somewhat precariously in a canoe under cover of darkness.

March-Phillips was one of those Commandos who stood out. One of his colleagues among the foundation group that included the likes of David Stirling, Roger Courtney, Paddy Mayne and Mike Calvert described him[2] as a 'gallant idealist and a strange quixotic genius'. He put together an equally quixotic team and acquired an excellent sturdy craft, a trawler named *Maid Honor*, which had been specially requisitioned for the purpose and would become the force's floating base, moored unobtrusively at Poole. However, a larger task suddenly appeared on the horizon, to spot and report U-boats operating off the west coast of Africa. The force sailed the trawler with a professional

[2] Peter Kemp, *No Colours, No Crest*, Cassell, 1958, p. 58.

skipper to Freetown and from that initial base began carrying out clandestine missions for the SOE. These missions have never been revealed,[3] although they must have been of some significance because March-Phillips was awarded a DSO on 24 July 1942, Appleyard received a bar to his MC and another of their party, Graham Hayes, also won an MC.

Mountbatten knew their story well and had met the *Maid Honor* team when they came home on leave in February 1942 and said then that they would make a damned good raiding party to be launched under his Combined Operations. The *Maid Honor* force thus returned to England in the summer – leaving the trawler from which they took their name in Lagos, eventually to be sold. Although officially still under the control of and financed by the SOE, Mountbatten took them over for some 'interesting little jobs' launched from home waters. They were now to be equipped with a motor torpedo boat, No. 344, which they would retain and make their own, and were given some marauding missions around the coast of France; they grabbed every opportunity. The first trial run was in fact a couple of nights before the raid on Dieppe, when they landed north of Cherbourg, killed three Germans, reconnoitred the surrounding countryside and returned home. The fact that they should have gone higher up the coast, to attack an anti-aircraft gun site at Barfleur but landed in the wrong place, was on this occasion overlooked.

Their next mission was on 3 September. They raided the lighthouse at Les Casquets, off the Channel Islands, with the object of capturing any personnel, documents and codebooks. Such places were always a good source for the latter, and this one was to be no exception. March-Phillips took 11 men on the MTB. They scaled a 25-metre cliff, rushed the lighthouse and found most of the occupants sleeping. They rounded up seven German soldiers and two telegraphists without a shot being fired and brought them back to Britain. 'A characteristic of those in bed,' the operations report noted, 'was the wearing of hairnets which caused the commander to mistake one of them for a woman.'

The party and prisoners were safely delivered back to base, where a good deal of useful information was gleaned from the raid. The success of the mission inspired others: first a reconnaissance landing at Ile Burhou in the Channel Islands on 8 September and then back to the French coast five nights later for a raid. They were to land close to the

[3] Although many of the papers relating to the Special Operations Executive have been released into the Public Records Office over the years, the *Maid Honor* force documents remain closed until 2017.

seaside village of Ste-Honorine on the western sector of the coast eventually chosen for the D-Day landings. It was to go badly wrong. They were to carry out reconnaissance and attack the first German stronghold they came across and attempt to bring back prisoners.

The landing site was shown on aerial photographs that clearly indicated a small German presence, but unfortunately on a moonless night the point preselected could not be identified and March-Phillips took the 344 close to Ste-Honorine's little harbour. Again, 11 of them went ashore using Goatley boats, while Geoffrey Appleyard, who had injured his ankle in a fall on the lighthouse raid, stayed on the MTB. Not long after the party landed, they ran straight into a German patrol. From the 344, lying up offshore, Appleyard heard machine-gun fire and then shouting, first March-Phillips's voice and then Graham Hayes's, and though he could not make it out clearly the message seemed to be telling him to make a run for it.

Even as he did so, the MTB came under heavy fire. Appleyard cut the anchor and sped out to sea, and when he returned again after a short while he saw that German patrol boats were setting off in his direction. There was nothing else for it but to abandon his 11 colleagues on the shore. A few days later, intelligence sources confirmed the worst news possible: that Gus March-Phillips was dead, drowned off the coast apparently trying to swim out to the MTB. Two others, Sergeant Williams and Private Leonard (or Lechniger, to give him his real name), were shot by the Germans. Seven of the remaining party were captured and made prisoners of war; two of these later escaped, the rest staying put for the remainder of the war. Graham Hayes had made his dash for freedom at the scene of the raid and escaped, eventually contacting local French Resistance agents through a prearranged method. After hiding him for several weeks, they sent him on a journey into Spain, where he believed he would be safe. Unfortunately, he was not. His identity was revealed by a traitor, and Hayes was arrested and subsequently handed over to the German authorities. He was taken into Germany, held in solitary confinement for nine months and shot before a firing squad in July 1943 for the crime of being a British Commando.

By that time, Hitler's edict to annihilate all British Commandos was being pursued to the letter and, ironically, it was to some extent the contribution of the surviving members of March-Phillips's SSRF that brought it about. Geoffrey Appleyard was given command of the force and, on the advice of Mountbatten, quickly built it back to strength and tried to put behind them the tragedy of the loss of so many comrades in one fell swoop. The way to do it, Mountbatten told him, was to get back into action straight away, so within a fortnight the SSRF was retraining under Appleyard for another raid. The team he put together was in many ways a milestone in the annals of small-party raiders. It included Second Lieutenant Anders Lassen, a Dane who was later to become a legendary figure in the SAS and a hero of the Royal Marine Commandos at Lake Comacchio (as we will later discover), and Captain Philip Pinkney, another future SAS star who became famous for his capacity to convert to instant food anything that crawled, hopped, walked and grew and made him an object of speculation even in the SAS.[4]

Pinkney was transferred from No. 12 Commando, along with four other volunteers, to the SSRF. They joined seven existing members of the unit for a raid on the island of Sark on the night of 4 October, still using the original MTB 344. They gave themselves a tough challenge, opening with the scaling of a 45-metre cliff climb. The sure-footed Lassen went on ahead to scout the scenery but found nothing exciting and sat by the track waiting for the group to join him. The group followed a footpath and tracks towards some cottages which proved to be uninhabited and moved on to a house thought to be a base for German soldiers.

They broke in through a window and discovered that far from enemy soldiers, the house was occupied by a widow who lived alone. Her name was Frances Pittard, and, they now discovered, she was the daughter of a former officer of the Royal Navy whose late husband had been the island's medical officer. For the next half-hour, while plying the grubby black-faced troops with coffee, she gave them a great deal of information about the Germans' activities and, in particular, where a number of them could be found at that moment. They had taken over the annexe of Sark's Dixcart Hotel, not a great distance from where the Commandos sat.

Although they offered to take her back to England with them, Frances declined because she said she had lived in the Channel Islands all her life and had no intention of leaving on account of a few Jerries. The SSRF troop padded off towards the hotel and prepared to make their entrance. The place was in darkness, not a soul about or on guard duty. The sleepy isle of Sark was not the most likely place for the sudden appearance of midnight marauders from across the water. They made a silent entrance, swiftly working through the unlit corridors, and counted five Germans fast asleep, so soundly, in fact, that the raiders nicked their pay books and weapons before rousing them from their slumbers with some scare-the-living-daylights tactics that resulted in all being herded outside in their nightshirts and pyjamas.

In the darkness, one of them made a dash to escape but was caught,

[4] Tony Geraghty, *Who Dares Wins*, Arms and Armour Press, 1980, p. 13.

then broke free again and was shot when he refused to stop. Three others, shouting and screaming, also attempted to escape in the melee. Two were shot immediately, and the third was 'accidentally' shot while attempts were being made to keep him quiet. The fifth German put his hands up high and kept them there. He was brought back to England as the SSRF made their escape.

Within two days, however, it became clear that one of the four they believed they had shot and killed on Sark had survived, and he had reported events of the night to his superiors. The German High Command immediately protested through the media that the British were at it again: they complained that the four soldiers on Sark were tied up and shot by British Commandos. The Germans went on to list a number of similar incidents and could only conclude that the Commando forces were trained and instructed in the art of tying up and killing their prisoners whom they did not want to be troubled with. The Germans claimed that the same barbarity had been seen at Dieppe and, earlier, in the Lofoten islands, when the Commandos, the Germans said, had bound and gagged prisoners and killed others who were coming to surrender. The heavy-handedness of the raiding Commandos at Lofoten was confirmed, it may be recalled, in the unconnected testimony of Bernard Davies in Chapter 3 in which he said 50 years or so later: 'We slaughtered a few Germans quite unnecessarily in my view; they would have surrendered anyhow.'

In other words, they were shot before they had the chance to put their hands up, a point also raised by Ronald Swayne, who was captured at St-Nazaire. He was among those prisoners of war shackled for a while, and the situation was discussed then, and after the war, and he insisted:

> Hitler started the image that we were ruthless thugs. It was totally false. The officers and men and NCOs were fairly hand-picked. We were most of us very fit physically, played games and that sort of thing, but enjoyed long hauls over the mountains. I think we were rather the opposite of thugs; certainly the officers were. A lot of the officers were a bit intellectual. A number of them wrote books later. They were anything but thugs. There was a difficulty, however, when you had very few men on unfriendly territory. If you captured a lot of enemy the problem is what to do with them. Whether any prisoners were ever killed by the Commandos after being taken prisoners, I really don't know. I know that on one raid, I think it was the Vaagso [Lofoten islands] raid, the soldiers were encouraged to shoot before the enemy could put their hands up. In other words, to try to prevent a lot of men being in a position to give themselves up,

> but I believe it was quite useless and they took enormous numbers of prisoners and had quite a problem.
>
> I personally never saw anybody behaving like that. I dare say it did happen. I mean, you obviously get some bad apples in any barrel. I think we would have regarded them as bad apples. It would have been quite foreign to my outlook, certainly, and I think all the people I served with, to behave like thugs. They [the Commandos] always went in pretty hard, of course, to create a shock.

In the immediate aftermath of the Sark incident, however, the whole business took another turn. On 9 October the German High Command, who painted a picture of their own troops being whiter than white in this regard – a blatant lie, of course – had henceforth decided to shackle 1,376 British prisoners positively identified as Commandos and not let them out of sight of guards 24 hours a day. Berlin radio announced that this was the result of 'barbarous and clumsy methods employed by the British military authorities'. The following day, the newspapers in Britain and Germany reported that the Red Cross was acting as a mediator and that hopes were high that the prisoners would soon be unfettered. Negotiations to achieve this were slow and then stopped altogether when word of the Sark raid reached Hitler's ears.

He threw one of his famous tantrums and dictated the edict, otherwise known as the infamous Commando Order, published in memo form on 18 October 1942, in which he demanded of his army commanders down the line that all Commandos taken prisoner from that day forth should be shot and all Commando units engaged in battle by German soldiers should be destroyed utterly and completely; no excuses would be tolerated.

Furthermore, the Germans carried out retaliation on Sark, deporting Frances Pittard and others to a place in France where they could be watched, thus causing some resentment in the Channel Islands about the Commando raiders who kept attempting to kill the Germans on their soil. The talks on unshackling the prisoners temporarily broke down, but the Red Cross managed to restart them and finally brokered an agreement whereby the prisoners in German camps would be released from their chains on 12 December on condition that the British provided assurances that they would issue an order forbidding any kind of fettering of prisoners, whether on the battlefield or in the prison camps.

Even so, the edict concerning the execution of Commando-style prisoners captured in German-held territory remained in force, as was soon to be demonstrated when Mountbatten, urged on by Churchill, launched a series of small-scale raids in that autumn of 1942, all timed to harry German positions as the Allies were launching their biggest troop-landing of the war, Operation Torch.

The edict remained a German secret for the time being. Had they known about it, the reaction of the Commandos might be termed in modern parlance as: Up Yours, Hitler. Far from being scaled down, the SSRF, which had started the renewed commotion over tied-up prisoners, was trebled in strength, and by the end of October six coastal motor boats were added to its establishment in addition to the battered MTB 344. Additional manpower was drawn from troops assigned on attachment, with three parties drawn from No. 12 Commando. It was still very much a Commando-style operation now to be commanded by Major Bill Stirling.

Like his brother David, who was tearing about the North African desert with his SAS, Bill Stirling was one of the original Commando stalwarts. He had already spent some time with the SOE and now returned to command the SSRF. Stirling, promoted to Lieutenant-Colonel, was to run his group side by side with another new outfit called J Force, recruited from regular troops from 102 RM Brigade, which was being established with headquarters at the Royal Yacht Squadron at Cowes, Isle of Wight, under Captain J. Hugh-Hallett, RN. He and Stirling were to work jointly on plans submitted direct to Mountbatten for approval. The first raid under the new umbrella, led by Peter Kemp, was to reconnoitre defences on the Brittany coast near Pointe de Louezec.

Difficult negotiation of a minefield led them into direct confrontation with German sentries soon after landing. One of the party hurled a grenade, and the result was described later by Kemp: 'Everything was obliterated by a vivid flash as a tremendous explosion shattered the silence of the night. From the sentries came the most terrible sounds I can ever remember . . . I felt a shock of horror that their soft, lazy, drawling voices which had floated across to us could have turned literally in a flash to such inhuman screams of pain and fear.'

As the sentry positions were raked with Tommy-gun fire, two more Germans ran out, rifles at the ready, to be shot down by the raiders. With Kemp's task interrupted now by German reinforcements, he waved the command to retreat and they dashed back to their boats and vanished in the darkness to return safely to base. A similar mission four nights later, using the newly recruited men from No. 12 Commando, was aborted by bad weather, which was increasingly affecting the cross-Channel sorties.

Meanwhile, another major raid, which involved air rather than amphibious delivery to the target, was in the final stages of preparation: Operation Freshman, a desperate – some said incredibly foolhardy –

mission that came straight from the Combined Operations desk of Mountbatten. Although not strictly a Commando operation in that it did not use personnel from any of the established units, it did involve training in their techniques, and Commando advisers were brought in to offer guidance on some of the daunting hurdles that the team would face.

For some months, British intelligence had been receiving a stream of good information from Norwegian agents that the Nazis were working flat out at a top-secret plant in Norway which was linked to German scientists' work on the production of an atomic bomb. One of the main ingredients was heavy water. Production of it was centred on the Norsk Hydro-Electric Plant at Rjukan, deep within the Telemark area and among the highest mountains in Norway, 60 miles west of Oslö and 80 miles from the coast. It was known that since the beginning of 1942 Berlin had ordered a threefold increase in production. The race was clearly on to get an atom bomb into production before the Allies, and at that moment in time the Germans were winning it.

Although the actual details were still sketchy, it was decided in London that the plant had to be put out of commission at all costs to cause maximum delay to the atomic project. But a veritable ring of steel provided by German troops, and a supposedly impregnable security system, along with a combination of natural hazards, made conventional attacks virtually impossible. The plant was in an exceedingly isolated position on one side of a valley densely covered with tall trees and undergrowth. Norwegian Resistance and British undercover agents had surveyed the region as best they could from the ground, and aerial reconnaissance had been carried out. From maps and photographs pawed over back in London, the size of the problem was evident.

The terrain on three sides rose up more than 760 metres from the river bed and lay in the shadow of Gaustal Fjell, 1,646 metres high. The plant itself was set on a rock shelf 305 metres above the floor of the valley. A single well-protected road, with progressive security checks, led to it. Bombing was out of the question, and the site was too far inland for an amphibious landing of troops because of the amount of explosives and other gear they would have to carry. The same difficulty applied for parachute troops. Only when the first heavy-duty Horsa gliders began rolling off the production line for Britain's newly formed Air-landing Brigade did the final plan emerge: Commando-style sabotage teams of troops would be delivered in gliders, towed by Halifax bombers and released as close as possible to the site.

Even that method was dangerous, with the chances of the selected men coming back judged to be on the wrong side of 60–40; in fact, it was virtually a suicide mission. Two possible landing zones were located by the agents about five hours' march away, and gradually the assault plan began to take shape. To set enough explosives to blow the place beyond repair was estimated to need a team of at least a dozen men, and with enough spare in reserve if – as was very likely – there were any casualties on landing. As the air landing of troops was still in its infancy in Britain, barely tried and tested, Mountbatten decided that they should send two teams, so that the whole plan was duplicated; if one didn't make it, perhaps the other might.

The men also needed to be skilled engineers, and they were selected from volunteers from the 9th Field Company (Airborne) Royal Engineers and 261st Field Park Company (Airborne) RE. They were under the command of Lieutenant A. C. Allen, RE, and Lieutenant D. A. Methuen, RE. Two Halifaxes, the only aircraft capable of towing a Horsa glider 400 miles and returning to base, were earmarked for the task under the command of Group Captain T. B. Cooper of 38 Wing, RAF.

The volunteers were put through a rigorous training routine under the guidance of Commando and parachute instructors. In the first week they went through weapons practice, map-reading and long marches. In the second week they were hauled off to North Wales for mountain climbing and camping out in dire weather on tinned rations. The third week was taken up with technical training, including a visit to Scottish heavy-water plants similar to the one in Norway. During the last week the group was kitted out with all their gear, staged a dress rehearsal and studied a scale model of the plant.

The air crews, meanwhile, needed to practise towing Horsa gliders. They were a totally new addition to the British military establishment, replacing the only other British gliders, which were much smaller and lighter. Even then, there was one more vastly unpredictable hurdle to overcome: the weather in mid-November was notoriously bad, and Norwegian meteorologist Lieutenant-Colonel Sverre Pettersen was brought to Scotland with his charts to predict the best day for the attack. They settled on 19 November, but in the event the skies were leaden and winds getting up as zero hour approached and Pettersen suggested a postponement. Coded radio messages from near the landing site, however, reported a relatively clear day, but there was always the possibility of a rapid change as the planes neared the mountains.

The air crews and the assault force agreed: they would move immediately. So at 1750 on Thursday 19 November the first of the two Halifaxes took off from Skitten airfield near Wick tugging the massive Horsa behind it, laden with 17 men and their gear. The second took off 20 minutes later. They were to maintain strict radio silence until they neared their approach, when a coded signal would be sent to base. In Norway, agents were waiting to guide them in using a Rebecca-Eureka air-to-ground radio-signalling system smuggled into the country a week earlier by Combined Operations.

Nothing more was heard from the aircraft for five and a half hours, when a faint signal was picked up from one of them asking for a course home. It had not managed to get close to the target landing area. Dense fog had closed in as they reached the coast, and then they hit a white-out: a snow cloud. The pilots searched as best they could for an opening. Fuel was running low, and then, with both tug and glider icing up, the tug hawser snapped and the glider sailed earthwards. The Halifax crew could do no more and set course for home. They arrived back in Scotland at 0151 on 20 November.

The glider crash-landed over the west coast, just north of Stavanger. Eight members of the team in the glider were killed outright. The dead were buried on the spot. The nine survivors were all captured. Four suffered various injuries and were taken to a military medical unit. The remaining five were taken to the local Gestapo headquarters, where they spent two months under interrogation and torture before they were shot on 18 January 1943. The four in hospital were, that same night, injected with morphine and air by a Gestapo doctor. The doses merely lapsed them into a semiconscious state, and in due course they were tortured and suffered horrific deaths. One soldier had been tortured – attached to a radiator with a leather strap around his neck – and eventually died after the Gestapo officers had lifted him up and down.

A leather strap was placed around the neck of another of the soldiers and his head was banged repeatedly on the floor. The third and one of the most seriously injured was left writhing in agony on the floor until he was killed by one of the Gestapo who stamped his foot on the neck of the dying man. The last survivor, less severely wounded, remained conscious until first light, when he was driven to a Gestapo base at Eiganesveien and there shot in the back of the head as he was descending cellar stairs. The following day all four bodies were collected by truck and driven to the quayside and dumped at sea.

Those shot by the Gestapo in January were buried at Trandum. Their bodies were recovered in August 1945, when they were laid to rest, with full military honours, at Vestre Gravlund in Oslo. After the war, three members of the German Gestapo, including the doctor involved in the deaths of the wounded soldiers, faced a war crimes court in Oslo. Two were sentenced to death and a third was given a life sentence. A Norwegian collaborator was also arrested but committed suicide before he could face trial. The doctor died before being brought to trial.

The second party almost reached its target, the Halifax flying close to the landing zone. Norwegian agents on the ground heard the plane's engines roaring in the clouds above them, but the radio-signalling beacon failed to operate and they could not give the crew a bearing. Nevertheless, the glider was released and crash-landed in the mountains near Helleland; the iced-up aircraft, after just clearing the mountain there, crashed into the next range. None of the RAF survived the crash. In the glider, three of the seventeen-man team were killed instantly. The rest, including four walking wounded, made a run for it and tried to enlist the help of a Norwegian official but were in fact handed over to the Germans as prisoners of war. The Norwegian was unaware (as were the prisoners themselves) that Commandos would never be treated as prisoners of war, and they were all shot a few hours later under the terms of Hitler's Commando Order.

The Norsk Hydro-Electric Plant was put out of action in January 1943 by a team of Norwegian saboteurs whose own heroic story was eventually fictionalised by Hollywood and became *The Heroes of Telemark* starring Kirk Douglas. The teams who volunteered for Operation Freshman became the forgotten heroes of an audacious scheme that, in the end, was let down by the failure of a small electronic beacon. It remains to this day one of the most gallant small operations of the war. Their task, taken up by a Norwegian sabotage group who followed them in, led the fight against a far greater threat than anything the world had ever seen. By the skin of their teeth the joint efforts of British and Norwegian troops prevented Hitler from grasping the potentially devastating power that could have turned the course of the war in his favour.

The Freshman teams were already posted as missing when the last of the raids run by Combined Operations in 1942 was underway. Again, it was an ambitious, over-optimistic and thoroughly dangerous project, by then codenamed Operation Frankton[5] but more popularly known as The Cockleshell Heroes through the later film of that name. The project was first mooted earlier in the year when Lord Selborne, Minister for Economic Warfare, wrote to Winston Churchill voicing his concerns about Axis merchant ships getting through the British blockade along the west of France. He wrote again in the summer, pointing out that the port of Bordeaux was especially busy, and intelligence reports showed that vital imports of rubber, tin, tungsten and vegetable oils were being off-loaded there for transshipment to Germany. In early September

[5] A full account of Operation Frankton, including Hasler's original plan and operation report, appears in *SBS: The Inside Story of the Special Boat Service*, Headline, 1997.

The locations and the technology may have changed but the Commando ethos remains the same, demonstrated in these photographs of events separated by almost 40 years: the D-Day landings (above) and the Falklands War. Then, as now, each man carries his self-sufficiency pack, weighing 100lb or more, on his long march into action. (*Royal Marines Museum, Eastney*)

Wounded are dragged clear at St Aubin, Normandy, and (below) the victors march on – as seen in this photo from the album of Ralf Haywood (X Troop, 46 RM Commando) pictured at shell-smashed Arromanches, 8 June 1944. (*Royal Marines Museum*)

(Above) Heading for one of the fiercest battles in the final stages of the Second World War, the Commandos face heavy fire from shore batteries at Walcheren, in November 1944, surrounded by stakes with mines attached. (*Crown copyright*)

Survivors of one of the hit and sunk landing craft at Walcheren are rescued by comrades. (*Royal Marines Museum*)

Final onslaught against the Japanese in the Far East: beach landings in small, timber craft at Chebuda and then (below) into the sweaty, danger-filled jungles of the Arakan, Burma, where booby traps and Jap ambushes abounded. (*Royal Marines Museum*)

Combined Operations planners produced a draft proposal for a raid on Bordeaux to sink enemy shipping. In the wake of the Dieppe disaster, the plan was rejected by the Chief of Staffs Committee on the grounds that the port, 62 miles inland, was too inaccessible.

Mountbatten did not give up, and later that month he produced for the CoSC a finely detailed plan drawn up by the commander of the Royal Marines Boom Patrol Detachment, Major 'Blondie' Hasler. Hasler, who had spent the summer months training his new Royal Marines Commando formation and was touting for work, proposed to lead a force of 12 men – himself included – who would use canoes to reach the port. They would be transported by submarine across the Bay of Biscay to a point close to the mouth of the river Gironde, 500 miles south of Plymouth. There, they would float off in six two-seater canoes packed with stores and explosives, paddle the 90 miles or so to the haven where enemy ships berthed, attach time-fused limpet mines to their hulls, sink the canoes and, with the help of the French Underground, escape back to England via Spain.

That, in a nutshell, was the plan. Mountbatten represented it to the CoSC on 30 October and enthusiastically announced that it had been 'planned especially to meet Lord Selborne's requirements' as set out in his several letters to the Prime Minister. This time, the proposal was approved and, it must be remembered, at that time the Allies were still unaware of Hitler's general order regarding the annihilation of Commandos. Few present at the meeting would have given the enterprise much chance of success, but with 'only a dozen men' at risk they were prepared to let it go on. The Frankton force, made up entirely of young volunteers from the Royal Marines, then began four weeks of intensive training for their epic canoe journey. Hasler himself drew up his list of stores and had canoes built so that they would carry a diverse range of food and equipment, including a set of civilian clothes and an explosives package that alone amounted to 48 limpet mines, 12 grenades, silenced Sten guns each with 3 magazines and weighing in total 180 kilogrammes.

The raid was duly launched on 5 December 1942 after intelligence reports confirmed that 12 large supply ships were presently berthed at Bordeaux. The 12 intrepid canoeists went aboard the submarine *Tuna* to be transported to the point off the mouth of the Gironde where they would disembark with their canoes and set off on the marathon paddle. On arrival at the given point, a periscope reconnaissance revealed numerous German patrol boats in the area, so disembarkation was delayed for 24 hours. The submarine surfaced again the next night when, apart from roaming searchlights on the shore visible under a 'beastly clear night', the coast was clear. Soon after eight o'clock, the crew was unloading the canoes into the water.

The men hit trouble even before they set off. One of the canoes was damaged while being unloaded from the submarine and was beyond repair. The party was now down to ten men. They took to a choppy sea, led by Hasler himself with his paddler, 22-year-old Ned Sparks. The journey would be slow and laborious, and despite a comprehensive study of tides they were hit by a swell that developed into steep rolling waves in the shallows as they tried to dodge the German searchlights. Barely had they entered the river itself than one of the boats was found to be missing. A second capsized soon afterwards after being caught in a tidal, leaving its two occupants attempting to swim for shore. The party had just gone beyond the mouth of the Gironde when, close to Le Verdon, they had to take action to avoid being spotted by three French ships anchored at a jetty inshore. On getting clear of this danger, Hasler discovered that a third canoe was missing. Back in Berlin, meanwhile, the Germans let it be known that they were aware of the attempt to sail up the Gironde, presumably after capturing the crew of the first canoe, and put out some nasty disinformation stating that the British sabotage party had been 'engaged at the mouth of the Gironde River and finished off in combat'.

In fact, the remaining two canoes were continuing with their journey fraught with danger and difficulties and at times requiring superhuman effort, laying up by day and travelling by night. They did indeed reach the port of Bordeaux on the night of 10–11 December but were unable to get to their targets until the following night. Working under cover of darkness and always in fear of being spotted, the two crews weaved between the ships and planted their cargo of limpets. Seven ships were attacked and were badly damaged when the explosives blew. Hasler had by then lost contact with the other crew members, and he and Sparks carried on with the plan as arranged. They scuttled their boat and made their way to a prearranged rendezvous with the French Resistance to begin a 1,400-mile journey home that would take them through France, over the Pyrenees into Spain and finally to the British territory of Gibraltar on 1 April 1943.

The remaining eight men were never heard from again. It was many months before news of those who vanished from the adventure, two by two, came through. The first, via Red Cross channels, confirmed that the two men in the capsized canoe had drowned. The fate of the remaining six was later discovered in documents found at the German High Command, Foreign Department/Security. A report by Major Reichel headed 'Sabotage attacks on German ships off Bordeaux', confirmed that 'a number of valuable German ships were badly damaged off Bordeaux by explosives below water level by British sabotage squads'. The report concluded with the chilling words:

All those captured were shot in accordance with orders on 23 March 1943.

Chapter Nine

The Duke of Wellington's Last Stand

After the Gironde incident, Commando raids on the French coast were put on hold. The action switched now to North Africa, the unhappy scene of the decimation of Layforce, as the Allies were finally poised to begin a new offensive in North Africa with the forthcoming invasions of the Vichy-French-held coast of Algeria before advancing on into Tunisia, leapfrogging on to Sicily and then into mainland Italy. Many of the Commando-linked units would be engaged in these operations over the coming months, and the organisation was already expanding and regrouping to meet the challenge. The Special Service Brigade, under the command of Brigadier Bob Laycock, by then consisted of ten army Commandos,[1] along with the Small Scale Raiding Force, known as No. 62 Commando, and an intelligence unit that went under the cover name of No. 30 Commando. An additional Royal Marines Commando was also formed: No. 41, emerging from the existing 1,000-strong 8th RM Battalion, which was reduced to 450 men, the remainder failing to pass the intense training and selection methods at the Achnacarry Commando Centre. They would take their place alongside 40 RM Commando.

Under the same umbrella of operations was 'Blondie' Hasler's RM Boom Patrol Detachment, which had a section bound for sabotage operations in the Mediterranean to blow up enemy shipping, and Roger Courtney's No. 2 SBS, which would send a group to North Africa on specific secret missions while retaining personnel for landing agents in France and projected reconnaissance missions in the Channel Islands. The new Combined Operations Pilotage Parties specialising in marking

[1] Nos. 1, 2, 3, 4, 5, 6, 9, 10, 12 and 14, each consisting of five troops of three officers and sixty-two men, plus a heavy weapons troop with mortars and heavy machine guns.

beach landing sites were training up new units to provide advance intelligence on all landing areas, as were the Royal Navy Beach Commandos, who would guide the troops in. In the Middle East, the Long Range Desert Group was still going strong while Stirling's SAS[2] and No. 1 SBS were assigned to support the North African invasion plans.

They were all variously to join the hard fight-back that was to put the Axis forces on the run. The push began on 30 October 1942 when General Montgomery, commanding the 8th Army, sent 1,000 guns thundering into action against the German's Afrika Korps at El Alamein to score his first famous victory. A week later, the second phase in the Allies' reclamation of the region, named Operation Torch, was underway with the largest ever assembly of Allied ships and aircraft for the invasion of the North African coast.

The British Commandos joined the party, although small in number compared with the overall strength of the landings – some 65,000 men in 670 ships, 1,000 landing craft[3] and a planned build-up to some 1,700 aircraft to attack a 900-mile front. While it is impossible to track the movements of these multifarious Commando groups in action, some of the highlights demonstrate their versatility in the theatres of heavy and often hand-to-hand combat in which they were now to become embroiled. Their involvement, at least in the early stages, was still blighted to some extent by misuse under army commanders who did not fully appreciate their skills or were simply short of Commando numbers in the line. The result was a rapid decline in manpower through casualties that might otherwise have been avoided.

Torch called for the arrival of 25,000 American troops who had sailed direct from their home ports in the US with 250 tanks to land around Casablanca. Another 18,500 with 180 tanks sailed from Britain via Gibraltar to land around Oran, Algeria, and it was intended that these two forces would combine to form the 5th Army. A joint force of 20,000 men would simultaneously secure Algiers and, as the 1st Army, it would then move swiftly to capture four key ports of Bône and Philippeville, in Algeria and Bizerta and Tunis, in Tunisia.

Nigel Clogstoun-Wilmott's Coppists[4] were in from the start, carrying

[2] Stirling was captured in January 1943 and spent the rest of the war in Colditz. His deputy, Major Paddy Mayne, took over the SAS, while Major Lord Jellicoe took charge of No. 1 SBS, which he renamed the Special Boat Squadron, attached to the SAS.

[3] The Royal Navy possessed only six such craft at the onset of the war.

[4] The top-secret group, the Combined Operations Pilotage Parties, had become known as COPPs, its members as Coppists.

out their first major pre-landing surveillance and mapping of beaches where the forces would come ashore. The Coppists' contribution was vital and would continue throughout the remainder of the war. Their tasks were ultra-dangerous, requiring strong swimmers, a cool nerve and good technical know-how in terms of plotting underwater hazards on all beach approaches and those on land that might hamper the advance inland or provide secure firing positions for a defending enemy force. Mountbatten himself, convinced by the arguments of Wilmott, fought to get approval for the COPPs' formation and in the end had virtually to demand support from service chiefs. The first recruits, only 18 in number, mostly officers recruited from the Royal Navy, the RNVR or SBS, went into training with little specialised equipment but were soon showing the way forward. Never again would the landings be based on such slack intelligence as occurred at Dieppe.

The same dangers were confronted by the expanding Royal Navy Commandos who followed the Coppists in once the landings became an operational 'go' to act as beach-masters, having to withstand the initial might of any defenders of enemy territory even before the assault force began to land. By the end of 1943, 22 RN Commandos had been formed, each consisting of parties of 20 men.

The landings in North Africa, followed by Sicily and the Italian mainland, would provide ample opportunity for all these new skills to be put to the test under the most gruelling conditions. Although the immediate strategic objective was northern Tunisia, the planners were already thinking well ahead, first to cut the Axis line of retreat from the advancing British 8th Army in the Western Desert. With Axis power then hopefully decimated along the southern shores of the Mediterranean, the way would be open to invade Europe through Sicily and Italy.

No. 6 Commando was the first of the Commando units to experience the war in real terms with the advance through Algeria and into Tunisia. They linked up with the 3rd Parachute Battalion to capture the town of Bône under a barrage of heavy round-the-clock air attacks. Their objective secured with few wounded, No. 6 then ironically took 43 casualties, with 11 killed, as they moved to a new position 60 miles west of Bône when their train was strafed by German fighters. Little more than a week later, they suffered another 82 losses when they joined the 8th Argylls and the 6th West Kents for the advance west, fighting for the first time as infantry instead of in their classic raiding role.

No. 1 Commando at least had the opportunity for their preferred style of amphibious operations along the coast, west of Bizerta, to hit the enemy flanks, but they, too, would soon be required to strengthen infantry positions in the advancing line as it stretched across an ever-widening front. There were tough battles ahead, and few appreciated that the Commandos were ill-equipped for such a role, lacking the appropriate firepower and with virtually no transport of their own. They travelled light, with all their possessions slung around their necks, and carried none of the backup of conventional forces. The battles they fought cost them dear, and with no trained reinforcements immediately available, the Commandos were soon dwindling in strength. No. 6, for example, was down to a mere 250 men when operating as infantry in the Medjez el Bab region south of Tunis, as they were faced with the defence of 75 square miles under constant enemy attack. Although the men held their ground, the unit took heavy further casualties, reducing their active force to less than 200. General Eisenhower recognised their contribution under the most harrowing conditions and ferocious scraps when, as the Tunisian campaign neared its conclusion, he wrote in congratulatory terms: 'You have exemplified those rugged, self-reliant qualities which the entire world associates with the very name Commando. Please transmit my appreciation to the officers and men of your command.'

As the Allied armies slowly began to push the Axis forces into the sea, the invasion of Sicily loomed ever closer and initial preparations were secretly underway soon after the start of the new year. Two COPP teams began clandestinely surveying the Sicilian beaches well ahead, in late February. The party, formed only a matter of weeks earlier, badly needed rehearsals but barely had time for them because their reports were required by mid-March. The weather was wintry cold and the suits used for lengthy swimming missions were ill-fitting and often leaked. They pressed on. They were taken aboard three submarines from Malta and set off for the Sicilian coast, 75 miles away, for night-time operations. The submarines would surface two miles off the coast after periscope surveillance.

If all was clear, the Coppists would set off in pairs in canoes to their designated beaches to a point 180 metres from the beach. The paddler would remain in the canoe, suitably camouflaged, and attempt to maintain a stable position, unnoticed, while the reconnaissance officer would slip into the water. He would be wearing a hefty suit of rubberised fabric, which was supposed to give him buoyancy and protect him from the cold. The suit had a built-in lifejacket that could be inflated by mouth and pockets laden with equipment, including: a .38 pistol, a fighting knife, an oil-immersed prismatic compass, sounding lead and line, beach gradient reel, an underwater writing-tablet with chinagraph pencil, twenty-four-hour emergency rations in case of separation, and two torches to home in on the canoe for the return.

The swimmers were to record every possible detail to assist invasion force beach-masters to bring to the shore landing craft bearing heavy transport, tanks and guns and men without undue mishap and to establish suitable sites for piers and breakwaters. The survey would take two hours or more, and the swimmer would return to the canoe, and then back to the rendezvous with the submarine. The Sicilian recce ran into trouble from the beginning. Survey sites were found to be heavily guarded, with sentries posted at around every 100 metres.

First, the leader of the COPPs' expedition, Lieutenant-Commander Norman Teacher, RN, failed to return to his canoe and was presumed dead or captured. The former proved to be the case. His paddler, Lieutenant Noel Cooper, an experienced canoeist who had been on Operation Torch as a beach-master, returned to the submarine rendezvous completely exhausted after a long search.

In spite of that, Cooper went out again with Captain G. W. Burbridge on 2 March. They did not return and were never seen again. On 3 March Lieutenant Bob Smith and Lieutenant D. Brand failed to return to their submarine in rough weather. They were feared dead, but in fact they had been thrown off course and, unable to locate their mother sub, simply set a course for Malta and kept on paddling – 75 miles back to base. They completed the journey in just over two days and landed in the Grand Harbour in Valletta, exhausted. Meanwhile, the reconnaissance mission around Sicily was still in trouble.

On 6 March Lieutenant A. Hart and Sub-Lieutenant E. Folder, also from the Middle East section, did not come back. On 7 March Lieutenant P. De Kock and Sub-Lieutenant A. Crossley failed to meet their connection, and the following night Lieutenant Davies went to look for them and did not return either. Others also went missing. Of the sixteen who joined the original mission, only four were known to be safe. Five – Teacher, Cooper, Burbridge, De Kock and Crossley – were never heard of again and were presumed drowned. The remainder had been captured.

The three lost officers of COPPs – Teacher, Cooper and Burbridge – were believed by some to have hit trouble and had taken the ultimate precaution against capture and torture by drowning themselves, thus not giving the Germans the chance to discover the purpose of their top-secret mission. Others disagreed and thought they had simply drowned or perhaps had been captured and shot and their bodies buried. In spite of the losses, however, the beach surveys went ahead with COPP reinforcements brought out to Egypt. Reconnaissance missions were completed successfully and would indeed prove their worth. Clogstoun Wilmott, meanwhile, had to get his force rapidly trained to cut the number of casualties on future tasks.

Even as the last German units were being booted out of the North African theatre, a massive armada of Allied ships, aircraft and soldiers was being assembled off Tunisia along with a vast stock of heavy-duty gliders for the first ever major air landing of troops into enemy-held territory. Three British parachute brigades and the new glider-borne 1st Air Landing Brigade would lead the invasion while Commandos were earmarked for their more traditional role as raiding parties to attack coastal batteries ahead of the major troop landings.

The capture of Sicily was planned as a rapid pincer movement. With Eisenhower in overall command, the US 7th Army under General Patton would land on the south-west coast while Montgomery's 8th Army would come in on the south-east. They would be followed in on the west side by the 1st Canadian Division and the 51st Highland Division, while the British 5th and 50th Divisions landed south of Syracuse. A number of coastal batteries were identified from aerial reconnaissance, particularly overlooking the beaches assigned to the Canadians and 5th Division.

The Sicilian landings were scheduled for mid-July, but even by the planning stage in March 1943 No. 1 and No. 6 Commandos had been in constant action since they landed five months earlier and were severely depleted and tired. There was no alternative but to withdraw them back to England to recoup and replenish. No. 3 Commando was at the time among the forces guarding Gibraltar, Britain's gateway to the Mediterranean, and they were brought out to begin training for Sicily on 6 April. In May they moved to Egypt to link up with the 5th Division, whose landing they would pre-empt with an attack on the batteries.

No. 2 Commando was also preparing to head out for Sicily, and seven RN Beach Commandos were already on their way. Bob Laycock himself also moved his own Special Service Brigade HQ staff to Egypt to oversee preparations as elements of his brigade when they began leapfrogging around the battle zones, conducting raids on specific targets such as coastal batteries or acting as a fire brigade to fill gaps in the Allied infantry lines in the face of strong German counterattacks. He brought with him Nos. 40 and 41 (RM) Commandos and elements of the 1st Canadian Division with whom the marine Commandos were to link up in the assault ashore, although there was little time for any real training.

Paddy Mayne's SAS, for the time being renamed Special Raiding Squadron, was shipped from North Africa to support the 51st Highland Division and had an initial task of capturing a lighthouse and gun battery on the cliffs at Cape Passero. In the countdown, all the Special Forces units were practising their landings and going over their timings, although in the end there was no accounting for mishaps or the weather which, on invasion day, 10 July, was unusually brisk for high summer, with gale-force gusts blowing up as H-hour approached. The paras were blown off course and the 300 officers and men of the 1st Parachute Brigade floated earthwards and landed in some disarray. In the high winds, the Horsa gliders laden with air-landing troops were buffeted, and inexperienced American pilots towing them behind Dakotas released half of them too early and 78 sparkling new Horsas plummeted into in the sea, with many casualties and chaps swimming for shore. Those who reached land made remarkable recovery and began the dash to their given targets.

The sea was running a heavy swell, and the landing craft were banging and crashing over the rollers. Many men from RM Commandos, experiencing their first action, were seriously seasick. As they headed towards their designated landing sites, searchlights swung in arcs across the sky, probing for the Allied bombers and fighters, which could be heard but not seen in the dark and cloudy skies above, preparing at any second to give the enemy a pummelling ahead of the landings. Then, as the aircraft let rip, the roar of dozens of very noisy landing craft engines echoed around the Sicilian beaches. The seasickness was soon forgotten, and as the men crouched behind sandbags, Tommy-guns ready, fingers on triggers, the engines cut to a low throb and the craft nosed forwards to within 45 metres of the sandy beaches, in accordance with COPP markers. The ramps crashed down and the men charged forwards, half expecting a barrage of gunfire, but in several of the landings none came, although machine-gun fire could be heard rattling away inland where the paras hit their first opposition. And so it went on through the night. The operation report on Paddy Mayne's squadron was fairly typical of all the Commando action:

> Squadron was lowered in LCA from HMS *Ulster Monarch* and *Dunera*. They touched down unopposed on the right beach at 0320 hours. The 3 inch mortars immediately engaged with the Btry with great accuracy. One troop was sent to stop any enemy reinforcements which might arrive from the west, one Trp to provide covering fire on the Btry while the third Trp carried out the assault. The Btry position was overrun without much difficulty by 0500 hours. Sixty prisoners were taken and some 50 killed or wounded. Four 6 inch guns, three Light AA guns, one range finder and several light and heavy Breda guns were captured. Our casualties were nil. Several airborne Div personnel, including Brigadier Hicks, joined up with the Squadron during the consolidation of the position. In view of the success of this operation Maj. Mayne decided to advance northwest to attack another 6 gun Btry some 2½ miles distant which had by now opened fire. The Squadron were therefore

> ordered to assemble at Dameria farm and at 0600 hrs continued to advance. The approach to the Btry was impeded by five or six defended farms met with en route. These were protected by MGs and rifles in addition to snipers' posts in concealed positions in neighbouring fields. The enemy were, however, attacked and knocked out in succession and a number of Airborne Div personnel, some of them badly wounded, were released from their captors. On arrival at the vicinity of the Btry the 3 inch mortars at once engaged the gun positions whilst Nos. I and 2 Trps attacked with fire and movement which came up against stronger opposition. Our casualties were however very light (one killed and two wounded) whereas we destroyed or captured five 80mm guns, one large range finder, four 4″ mortars and several MGs and LMGs. Major Mayne now decided to assault yet a third Btry at Punta Della Mopla. He therefore ordered No. I Trp to proceed to the assault while the 3 inch mortars engaged a fourth Btry No.1 Trp successfully accomplished its task capturing and destroying the following equipment: Three C.D. [Coastal Defence] guns (6 inch), one 40mm AA gun, two light AA guns, one range finder, several MGs and LMGs and the Bty Commander and all his personnel. The 3 inch mortars had in the meantime engaged the Fourth Bty with success, setting alight to the magazine and blowing up the ammunition . . . on completion of their allotted tasks, the Squadron joined hands with 5 Div and finally bivouacked at 2100hrs. Summary of casualties – enemy: 500 prisoners, 200 killed or wounded. Own: 1 killed 2 wounded.

Other units were not so fortunate in terms of casualties, especially the paras and the Commandos, although there was no doubt that the push through Sicily went along far more speedily than they had dared hope. There were still major pockets of resistance from the remaining German units, especially as they were pushed back towards the south and east of the island. No. 3 Commando, having successfully completed the first stage of its work with fewer than a dozen casualties, re-embarked on the *Prince Albert* on 13 July to land north of Augusta, move ashore and take and hold a road bridge over the River Lentini ahead of the advance of the 50th Division, at the time attacking the nearby town, and was expected to reach the bridge at dawn. The Commandos came under mortar fire as they made their landings from about 275 metres out but nevertheless quickly established a bridgehead and managed to get all men ashore with fewer than 20 casualties. They then made off towards the bridge and engaged a number of German or Italian positions en route. At the bridge, which they captured successfully, they discovered that the Germans had laid time-fused explosives, and they were set to blow in 30 minutes; they were defused by the Commando demolitions men, and then they set up their PIATs to keep the enemy at bay and waited for the arrival of the 50th. But then the enemy rolled up with a heavy tank and started pounding No. 3, who had no weapons capable of taking it on. Mortar and heavy machine-gun fire, too, was increasing, and at dawn there was still no sign of the backup force. Commanding Officer Lieutenant-Colonel Durnford-Slater had no option but to withdraw to the hills 730 metres away.

The enemy formations gave chase and had them pinned down by mortar and 88-millimetre guns with airburst shells that caused severe shrapnel wounds. By now, the casualties were increasing, with 60 killed or wounded and still no sign of the 50th Division. Durnford-Slater ordered his men to split and scatter and reassemble at a given rendezvous. Although they made a spirited fight-back, No. 3 Commando suffered 153 casualties, killed, wounded or missing in the action before the 50th eventually arrived to rescue them. They had, however, saved a vital bridge from being blown sky-high, and Montgomery stated that henceforth it would be called the 'No. 3 Commando Bridge'. An almost identical operation, and heavy engagement, was carried out by the 1st Parachute Brigade at Primasole Bridge and suffered a similar fate in terms of losses in a two-day battle, although again they saved it from being blown. Primasole became a battle honour for both the Parachute Regiment and the Durhams of the 50th Division who went to the paras' aid.

The two Royal Marine Commandos, meanwhile, had achieved an exemplary advance from their landing positions, although not without overcoming some substantial pockets of resistance. Ironically, their most serious setback in terms of losses came not during their encounters with the Germans in the field but when they were on board ship. They had joined Paddy Mayne's Special Raiding Squadron to go forward to Catania, but the advance was aborted at the last minute when intelligence reports confirmed that the Germans were heavily reinforcing the area. During the night, however, the RM Commandos' ship, the *Queen Emma*, was hit during an air bombardment, killing 16 men and wounding another 63. It was among the worst multi-casualty attacks suffered by the two Marine Commandos in the Sicilian campaign.

On 22 July Commando reinforcements arrived in the form of No. 2 Commando under Lieutenant-Colonel Jack Churchill, raconteur and rebel, who fought wars his way and who brought with him his customary eccentricity, which could be relied on to upset the top brass and bring a touch of humour to the darkest hour. He marched forwards from the landing at Catania, where his group were to be bivouacked, with his trademark claymore slung around his waist, a longbow with arrows (he was in the 1939 world championships) round his neck and his bagpipes under his arm, all of which were in constant use. His brother, Lieutenant-Colonel Tom Churchill, was already in Sicily, with Laycock as a staff officer.

Jack Churchill himself recalled some unusual developments as the campaign moved into its final mopping-up of German opposition on the island and began preparations for the advance into Italy:

> When No. 2 got to Sicily, we had a rough time for a while and took quite a few casualties. When it subsided and when we weren't actually fighting, we did live in better houses, which was entirely due to Randolph Churchill, who arrived to join Bob Laycock at his HQ. Winston, of course, used Randolph as a spy. He attached himself to the Commandos quite a lot. He got on very well with Laycock, and he sometimes came with me, same name and so on; we all got on very well with him. He was always flitting about, but he was of no great value. Bob Laycock at one time tried to incorporate him permanently into the Commando, but he refused, I think because he was also earning some money writing accounts of the war. Anyhow, there we all were in Sicily, and Randolph took one look at the brigade mess and complained loudly: 'This place is like a bloody pigsty. Do you really expect me to eat in here? Here we are, all Commandos, and we're not living any better than an infantry company. It's bloody ridiculous.'
>
> Bob Laycock said: 'Well, all right, if you think you can, do something about it.'
>
> So Randolph went off in his usual blustery way, got hold of the local mayor and said: 'I've got General Laycock here. What's the best house in the town?'
>
> Came the reply: 'Prince Borghesi lives there.'
>
> 'Chuck him out,' said Randolph. 'The British army is going to move in now.' And it was all fixed. Of course, in no time Randolph had seen Borghesi and said to him: 'I'm the son of the British Prime Minister.' He traded on it quite wilfully, particularly with the Americans; he had a wonderful entrée with them, and he flew all over the place. Laycock liked him and we, incidentally, liked Laycock. He was the perfect brigadier to have: he left you entirely on your own, and he assumed everything was going all right unless something goes wrong and then they have a hell of a witch-hunt.

A successful conclusion to the Sicilian campaign was already in sight. The Italians had no heart for battle now. Mussolini, who had ruled Italy for 21 years, was deposed on 25 July and King Victor Emmanuel assumed control. The anti-Fascist Marshal Badoglio became Prime Minister and chief of the Italian Armed Forces. The word was that Italy was pulling out of the war. The Allies had so many Italian prisoners they could not cope with the numbers, and in one famous incident when a British unit was heading into attack a crowd of Italian soldiers came out of a barn with their hands up and were simply told to bugger off. The Germans, however, still held a number of strongpoints, covering the withdrawal of their troops and equipment to the mainland, and fierce battles were still being encountered by the main invasion force, particularly on the plains and hills around Catania.

In the first week of August, Eisenhower called a planning conference in Algiers to outline his proposals to move into Italy. Bob Laycock and Tom Churchill joined other brigade commanders who flew for the briefing on the next move. Initially, the plan was for a straightforward invasion into the toe of Italy, but the top brass decided they wanted to speed things up and an additional landing would be made in the bay of Salerno, 30 miles south of Naples. This was confirmed in mid-August as the British and American troops met up in Messina as the German's final defensive position on the island collapsed and the whole of Sicily was now in Allied hands.

By then, long-range guns and Allied aircraft were already pounding coastal batteries on the mainland of Italy, and the Germans were retreating well inland to form a defensive line ahead of the Allied landings. The attack was set for 3 September – the day, in fact, on which Italy formally surrendered and decreed that its troops should lay down their arms. The Germans, in response, immediately piled in reinforcements to halt this vital Allied advance towards central Europe. Aided by the difficult Italian terrain and mountainous spine, the German planners threw everything they had available into Italy, which in the end amounted to sixteen well-equipped divisions, eight of which were under the command of Rommel himself, who was to mastermind the defence.

In the days prior to the landings, No. 3 Commando and the Special Raiding Squadron were commissioned to send parties across the three-mile Strait of Messina to report on enemy defences. They came back with the impression that they were light. Six patrols from No. 3 were sent in again on 28 August, this time with orders to remain and report back. When nothing was heard from the patrols, a search party went out to discover what had happened. It turned out that their radios refused to work. They then came under attack and scattered into the hills. One patrol was captured, although more than forty of the original party eventually returned to the fold when the main force landed five days later.

Montgomery's 8th Army went in across the Strait of Messina to land on the toe, with the remainder of No. 3 Commando, the SRS and No. 40 RM Commando coming in behind. The last was subject to some skirmishes when it ran into enemy gun positions, suffering more than 30 casualties. The Commandos sent out patrols to discover their positions, and Kittyhawk bombers were called in to put them out of action. The landings, however, were successfully completed within forty-eight hours, and then began a hard-fought six-month campaign to reach the north.

The American 5th Army and the British X Corps, along with No. 2 Commando and 41 RM Commando, embarked on a fleet of vessels to make their entrance on 8 September. Laycock and his headquarters came in with the Commandos, and they were to be confronted by enemy defensive positions in the hills surrounding the wide sandy beaches of the Gulf of Salerno. Jack Churchill's description, however, was more reminiscent of a voyage on the Grand Tour:

> We'd been resting at Catania after some heavy fighting in the final throes of the Sicilian campaign when I heard the news that we were to support a bridgehead at Salerno. My Commando was going in with 41 RM, and we sailed up the coast in the evening. It was quite a pleasant journey, rather civilised, in fact. Randolph Churchill, my brother Tom and myself were sitting drinking port as we came in to land at Salerno. We were just chatting, generally surveying the scenery, and Tom said: 'I'm always interested in the codenames they give to operations, and I wonder why in the heat of the summer here they've called this operation Avalanche.' Randolph said: 'I expect it's because there's an avalanche of Churchill's landing.'
>
> We were to take up position in Vietri, east of Salerno. It was a typical Commando operation. We were to take the batteries first of all, hold that flank and the Marine Commando would come in after us, and they were to go to La Molina defile and occupy the heights on each side in case the Germans came along. In theory, the army could then go through and take Naples. We landed without being shot at and then rushed up and sorted the battery out, which wasn't a problem. The Italians had just come out of the war, and the Germans hadn't really got themselves established, although it didn't take them long. We took the town and held it and gathered up a lot of prisoners.
>
> I established my own HQ in the town, although the bridgehead at Salerno was never increased; it was one of the few bridgeheads which was damned nearly lost. The Germans were too strong at the back and in the hills. They had an OP and the guns behind were

shelling everybody. They launched an attack against us from the east of Salerno and a lot of heavy fighting began there.

For No. 41 RM Commando it was what one of the operational reports described as 'a lively first two days, the hardest that 41 had ever had; losses were heavy'. As they came off the landing beach and up into the hills behind, they were attacked by crack German paratroopers and Panzers, and their commanding officer, Lieutenant-Colonel B. J. D. Lumsden, was wounded in a direct hit on his HQ. P Troop lost three-quarters of its men and Q had only two men standing at the end of the second day. They were pushed off the bridge they were defending and it took a brigade to reclaim it.

Another battle raged for control of the pass to La Molina, halting the advance until a counterattack was launched, spearheaded by No. 41 and No. 2 Commandos, forcing the Germans out. They then moved to a series of hill features, one of which had a German observation post on the top of it and No. 41 and No. 2 were given the task of clearing it. Jack Churchill led the way up:

> We needed to dispose of the OP and, once achieved, the main force coming in behind us would start to flow through to take over our positions. It didn't quite work out like that, and it was a bloody hard battle to dislodge them and we [the Commandos] were getting a bit annoyed because we were being used for attack after attack, doing things the army ought to have been damned well doing themselves.
>
> On the thirteenth there was a terrible battle at Dragonea, after which we handed over to the Leicesters. On the fourteenth we came down and we were asked to do a raid up a very steep incline to the road which led out of the valley to the top of a ridge, which the Germans had infiltrated and they had driven the Queen's Regiment into a tight corner. We had already suffered very bad losses but the general said: 'If you could do a raid up here this evening straight away, it will help us a lot.'

Churchill himself led the raid, with claymore in hand and with an NCO at his side. Between them, they captured 45 prisoners before the rest of the unit was signalled to come forward and take the village of Pigoletti. That task completed, Churchill led his men and prisoners back down, with a number of wounded carried on carts with huge wheels which, he commented, looked like something out of the Napoleonic Wars. It was manhandled by German prisoners. Three men had to be left behind, because he did not think they would last the tortuous journey down. As they neared the bottom of the feature, he told his men to go and get a meal and made contact with divisional HQ. They were, said Churchill, ecstatic, and told them not to come back down but to stay in position.

> I told them we were already down, but they told me to get everyone back up there fast. So No. 2 Commando went back up to Pigoletti and we strung out to the left where they had been originally, then 41 Commando came up on our right, extended the line to the right. We had Pigoletti and this horseshoe and a hill called the Pimple, and the Germans were now very close by there banging away at us with 88-millimetre guns. Now on the early morning of the sixteenth, my birthday, there was a big counterattack and we were badly hit. We held, all right, but lost quite a few people, killed and wounded. Then there was a second counterattack, and I wasn't too happy about No. 2 Troop, which was Wellington's troop [commanded by Captain the Duke of Wellington, known to all as Morny, from an earlier title he used, Lord Mornington]. I tried to get Wellington on the wireless – he was only about 275 metres away. My HQ was always practically in the front line where I could almost shout at people fairly easily. Morny was a very humble, quiet chap, for a person with a title, particularly a duke's title. Most people with titles I have found are tiresome, but he was anything but; he wasn't the slightest bit conceited.
>
> So I got on to the wireless operator there and said: 'Get Captain the Duke of Wellington.' And he said: 'He can't come to the telephone, sir.' I said: 'Why not, is he wounded?' He said: 'No, he's not wounded, but none of us can move . . . we're being attacked hard. I think we might be overrun.' I told him to call out to the duke and tell him that the colonel's ordered him to retreat, to come in. So about 20 minutes later, 2 Troop came back, and I said: 'Oh, thank goodness you're home, Morny. I gather you've had a very ghastly time.'
>
> He said: 'It wasn't too bad.' And it turns out his man on the radio was a bit of an alarmist. So Morny volunteered to go back because there would now be a gap in the line. He stopped for a cup of tea and then took his troop back. Now the Germans had moved up a bit during the time and almost immediately the Duke of Wellington was shot by a German officer. In fact, it later transpired that they shot each other at the same time, were both killed and were buried in the same shallow grave.
>
> The next morning, I rang Laycock's HQ, spoke to my brother first of all and then to Bob, and I said: 'We had a bad night and

early morning, brigadier. We had considerable losses, including, I'm sorry to say, the Duke of Wellington.'

'Oh, Christ,' he said, 'was he killed?'

'Yes, he was killed,' I said.

'Have you got his body?' he said.

I said: 'No, we haven't got anybody's body. The dead were buried in the fields. His chaps buried him straight away.'

He said: 'How do you know he was killed?'

I said: 'Well, everybody has told me and it's clear he was killed.'

So he said: 'Well, I'm not happy about showing a duke as killed. I'll show him as missing.'

I said: 'Well, look, brigadier, if you do that you'll give his mum, the duchess, and his sister Anne the impression that he might reappear when in fact he won't, he's as dead as a doornail.'

He said: 'How do you know, if you haven't got his body?'

And I said: 'Well, I told you, old boy, I'm absolutely positive. I have eyewitnesses.'

Laycock decided to lie and said: 'No, I'll post him as missing. I'll look bloody stupid if he turns up a prisoner two months later.'

So I said: 'Right, brigadier, but I'm going to write to his mum, the duchess, and his sister and say that he's dead but that you're showing him as missing for these reasons, but that I'm certain that he's dead.'

And he said: 'Right. If you like to do that, that's OK. You're his commanding officer.'

So that was quite agreed, and I wrote to both of them, but it was by no means the end of the story.

The next day a full infantry brigade moved in behind the Commandos and took over the fight, and the Germans were beaten off, finally withdrawing from the Pimple on the nineteenth. The Commando units had fought a brave battle for almost ten days and had suffered heavily. They were withdrawn to Sicily for a rest. Of the 738 Commandos who went ashore for Operation Avalanche, only 372 made it back to Sicily. They were losses that no one could withstand and clearly the lightly armed Commandos had been misused again – as attacking forces instead of raiders. They could easily do both, as Jack Churchill kept on pointing out, provided they were given the hardware appropriate for the task at hand. With some of the heavy stuff the Germans had been rolling in, they might just as well have been hitting back with peashooters.

In Sicily, also, the saga of the Duke of Wellington was resumed, as Churchill explained:

> I'd not heard from anyone who wanted to know the details of the duke's death, not a single squeak. But then, a curious coincidence: Morny's uncle, Gerald Wellesley, who was, in fact, also his heir because Morny had no brothers or children, happened to be in Sicily. He was in AMGOT – Allied Military Government in Occupied Territories, or better known as Ancient Military Gentlemen On Tour. Morny's side of the family hadn't spoken to the Wellesleys for years, even though the title would go to them unless Morny produced an heir, which he didn't.
>
> Now, when I brought No. 2 back to Sicily, Wellesley turns up and I had the displeasure to inform him that he was now the Duke of Wellington. The story didn't end there, either, because Bob Laycock became rather a nuisance about the duke's body when he heard of Wellesley's presence. He wanted me to retrieve it and positively identify that it was Morny who had been killed so as not to embarrass the man now claiming the title if the real Duke of Wellington turned up alive and well and living in Colditz or somewhere.
>
> Of course, I knew that he wouldn't, but, anyhow, I took a party of body snatchers [the body recovery unit] with their brown paper bags back to Salerno, dug up the graves at the battlefield where Morny was killed and found the decomposing bodies, one of which we were able to positively identify as that of the duke by a dental plate in his mouth and other things, such as his pipe and a ring. I said, 'This one is Wellington,' and they marked his bag, and I said, 'This one is a Hun officer,' and they marked his bag. Thus ended the saga of the Duke of Wellington. The other interesting thing that happened that morning: we got some mail which had been to about five different places and finally found its way to me, and the first letter I opened was the one saying that I'd just been elected to the MCC.

The cricket would have to wait. Although Naples was taken on 1 October, some long, hard battles lay ahead.

Chapter Ten

Send for the Marines? Not Bloody Likely!

As the Allied armies in Italy began their hard fight north, developments back in London were already focusing on the opening of the second front in western Europe in the shape of Operation Overlord and the D-Day landings. At the very forefront of these plans were the military groups whose formation Winston Churchill had demanded when he became Prime Minister in May 1940: the parachute and Commando brigades. His determination to see the creation of a 'force of 10,000 men' who could storm into battle by land, sea or air and rapidly attack targets that were beyond the reach of conventional forces had proved its worth. The special and irregular forces and all the little private armies drawn together in the face of opposition from many of the dowdy military leaders who did not possess that same foresight had shown their mettle in every theatre of the war. Now, they were to spearhead the greatest and most dangerous adventure of all.

Just as German storm troops led the Blitzkrieg across Europe, parachute, air-landing and Commando formations would dash ahead of the great herd that would soon be assembled to make its assault on the French coast to begin the first stage of the rout of the Nazis.

Although both the paras and the Commandos had originally been raised as storm troopers for specifically targeted assaults in which, lightly armed, they would do their business and move on, the needs of the war planners were changing as the Allies began to launch their concerted attacks on all German-held territory. As head of the Commando organisation, Bob Laycock knew better than anyone the capabilities of his forces, and in the early summer of 1943 he had produced a paper that set out the changes he felt were necessary to put the Commandos on a different footing. He surmised that the days of small-party raiding designed to harry and annoy the enemy were coming to an end. There were still opportunities for such raids, but they could safely be left in the hands of smaller groups.

Laycock foresaw a much wider brief for his Commandos, and it had as much to do with survival rates among his men as anything. Instead of being used as raiders, both the paras and the Commandos were being increasingly called on to fight in prolonged situations after they had already knocked hard on the front door and let everyone else in. Army commanders kept them *in situ* for long and arduous battles for which they were not equipped either in firepower or support. They travelled light, had no base, seldom had any artillery backup and carried everything they possessed on their backs with no transport. Commandos travelled in cramped ships quite often in dire conditions, with no facilities for off-loading any heavy equipment and went ashore in a variety of landing craft and small boats that quite often were barely serviceable.

Lately, as in North Africa, Sicily and Italy, their losses were high, and this highlighted another problem: there were no trained reinforcements immediately available overseas, and a Commando which may have lost more than half its men could be topped up only by drafts sent out from England or calling for untrained volunteers from the line. But there really was no substitute for the heavy-duty training at Achnacarry and elsewhere that instilled not only the methods of storming a target but basic life-savers, such as endurance swimming, route marching with a 45-kilogramme load strapped to their backs, fast reaction to a given event, silent approaches (surprise attack) or loud approaches (screaming and shouting to scare the living daylights out of the enemy), understanding and recognising all types of explosives, survival techniques if stranded behind enemy lines and the craft of listening for sounds unrelated to the countryside which may spell danger of another sort. These elements demanded weeks of training, and once a raid or target had been identified the work-up for the job would be precise and carefully timed. The intensity of the training was the reason why there had always been so many rejects. Men who did not make the grade for one reason or another were not only a danger to themselves but could put the whole troop at risk. The esprit de corps of the Commandos who had undergone collective training was a vital ingredient to their success.

Laycock foresaw these problems, and in his paper for Mountbatten he suggested that there were two alternatives: either to disband the Commandos or to put them on a surer footing, with a substantial enlargement of their overall numbers, a holding operation for reinforcements and at the same time an upgrade of their firepower. Support for the latter alternative came when the Anglo-American Joint Staffs under

General Frederick Morgan began to draw up their proposals for Operation Overlord. A draft plan was flown to Churchill while he was taking a break in Marrakech, where he was joined by both Montgomery and Eisenhower. When the two generals were shown a copy, Montgomery said he needed more 'initial punch'.

Since this would entail a spearhead force of paras and Commandos, it had already been the subject of some debate. Sir Alan Brooke, Chief of the Imperial General Staff, had called for nine army Commandos to be operational by late September 1943 and double that number by the time of the projected launch of Overlord. From a training perspective, that was a challenging target. Marshalling suitable manpower would be virtually impossible – until, that is, Mountbatten suggested that the Royal Marines should provide all the additional units. In fact, he went further – that the marines should launch a takeover bid for the whole Commando organisation.

They had already provided two Commandos and had plenty of capacity for more. That suited the marines. They had so far been given surprisingly little to do in the war and were getting very fed up. It was also pointed out that the 1922 Madden Committee report had suggested that the marines were a logical home for Commando-style amphibious raiders, yet that recommendation had never been taken up. With pressure on the manning of army units, Mountbatten's Combined Operations could see no obvious reason why the marines should not become the predominant Commando force, to which the army replied in very vocal terms: Not Bloody Likely.

However, in the late summer of 1943 it became a *fait accompli.* The marines were chosen to provide six more Commandos, to be known as Nos. 42 to 47 RM Commando, in addition to the two already in existence. A ninth, No. 48 RM Commando, would be added in the spring of 1944 in the build-up to Overlord.

It was a significant expansion of the whole Commando operation, and in real terms a complete change of direction and manpower. They would also get more firepower, their own engineers, a transport unit, medics and signals – all tasks previously handled by the men themselves virtually at Troop level. The army Commandos were not at all pleased, and there was considerable dissension over what many of the pioneers of the movement considered was an unwarranted and unwelcome incursion. The Commandos, they insisted, were shock troops *par excellence*, and untried and untrained marines simply would not do. The marines countered that they had been doing it since Elizabethan times, thank you very much, and no Johnny-come-lately organisation could take that away from them. Their colonels would say that theirs was a unique and proud history with a lineage that could be traced back to 1664, while conversely the army Commandos' very short history was not at all above criticism. But because the Commando idea had been 'Winston's baby', little had been said against them. A taste of the animosity was recalled by Brigadier (Lord) Lovat:

> I attended a conference to meet the marines in the summer of '43 and gave our views on the amalgamations. It was not a very tactful discussion. Some (Jumbo Leicester, David Fellows and their commander, Major General Robert Sturges) became good friends. The Royal Marine colonels, collectively, were not so amiable. Their outlook was, understandably, stiffer and more hidebound than our own. Great traditions lay behind them. We had none. They could drill and counter-march better than most of the regular army . . . but few of these marines had manned gun turrets or seen action in the days of capital ships. None had Commando battle experience or knew what close fighting was about. I personally found the patronising 'old pro' attitude to us 'upstarts' very hard to swallow.

But swallow he had to, because the amalgamation was going ahead, and as news filtered down the line men in the ranks were spitting with rage that they, all volunteers and proud of it, were now to be joined by conscripts and regular soldiers. One of the pioneers of the army Commandos, John Durnford-Slater, summed up the feeling of many of his officers and men: 'We were dead against the idea. I was proud of what we stood for. In spite of Mountbatten, whose idea this was, I felt convinced that units of conscripted marines could not be expected to maintain the high standard of shock troops.'

Later, however, observing No. 45 RM Commando in training with Nos. 4 and 6 army Commandos, he somewhat grudgingly conceded that they would probably make the grade 'in such good company'. There was merit on both sides of the argument, and it was the marines' commander, Major-General Sturges, who – seeing that the animosity might get out of hand – suggested that the two groups should be moulded together in new brigades rather than kept apart, so that each could work with the other, as they had already demonstrated in Italy. Even so, the resentment was not completely eradicated and continued among the veterans after the war when, as will be seen, it was the army Commandos who got the axe while the marines stayed in business.

For the time being, however, they were to be as one, with the army and marines mixed and matched together to form four Special Service Brigades, although the command structure and their groupings would pass through a series of confusing changes as the demands on their services developed with the progression of the war across all the theatres from western Europe, Germany, Italy, the Balkans and on into south-east Asia.

However, further changes in the hierarchy helped calm the ire. Bob Laycock, the man who walked across North Africa for 41 days after the failed attempt to assassinate Rommel and who had the respect of both his own army Commandos and the marines alike, was being moved back to London. He was to succeed Mountbatten who, at the age of 43, had been promoted to Supreme Commander of Allied Forces in south-east Asia. Churchill steamrollered the appointment against the same murmurings of dissent that accompanied his selection as Chief of Combined Operations. Mountbatten's number-one task in opening this brand-new theatre was the recapture of Burma from the Japanese. The humiliating retreat of the Allied forces into India a year earlier at a cost of 13,463 casualties had only recently been salved by the Commando-style operations of Major-General Orde Wingate and his Chindits behind enemy lines. Wingate, another swashbuckler who ignored conventional military thinking, had won the reputation as a latter-day T. E. Lawrence with the exploits of his guerrilla force, formed in February 1943. Ex-Commando 'Mad Mike' Calvert had also distinguished himself and had been promoted to brigadier in the Chindit operations. Wingate himself would not survive to see the full results of his endeavours. He was killed in an air crash in March 1944, and his body now lies in the Arlington Cemetery, Washington.

Mountbatten flew out to his new headquarters in India in October 1943. Before long, he would be calling on the assistance of the Commandos, Coppists and the Special Boat Service, although even as he left they were all already heavily tasked in the ongoing battles in Italy and in training for the upcoming big event: D-Day.

At the time of his appointment, Laycock handed over to his number two, Brigadier Tom Churchill. At the time, he had four Commandos and the Special Raiding Squadron under command in the No. 2 Special Service Brigade in Italy, while almost all their colleagues in the remaining units back in England would be retained there to spearhead the D-Day landings. We pick up the story of the Italian campaign in October 1943, when No. 2 Commando, headed by Churchill's brother Jack, and No. 41 RM had been resting after their heavy fighting around Salerno. Meanwhile, No. 3 and 40 RM Commandos, along with Paddy Mayne's SRS, went back into the fray. Their task was to land at the port of Termoli on the east coast of Italy and establish forward positions in advance of the arrival of the 8th Army.

Both Commandos were still under pressure in terms of numbers: No. 3 had received no reinforcements since the casualties taken during their landing on the mainland and was down to fewer than 200 men, while No. 40 under the much-respected marine Lieutenant-Colonel 'Pops' Manners could muster fewer than 400. No. 3 came ashore at Termoli safe and dry, but No. 40 hit a sandbank in heavy seas and had to wade ashore at a depth of 1.5 metres in which their radios were drowned. No. 3, meanwhile, had set about its task of forming a bridgehead a mile west of Termoli, while No. 40 and the SRS carried on through to capture the town and the main road out which led cross-country to Naples. They were to take and hold these forward positions about 15 miles in front of the 8th Army and remain there until the main force came through.

Lieutenant-Colonel Peter Young, wounded in earlier skirmishes, discharged himself from hospital to join his No. 3 in the anticipated scrap at Termoli. It came at dawn the day after they moved into position, when fighter bombers swooped low and gave them a pounding before the German ground forces made their counterattack, soon to be backed up by the unexpected arrival of the 26th Panzer Division, which had been resting after the action in Salerno. The Panzers had driven from west to east and now rolled up with infantry and tanks and started blasting the Allied positions. The fighting developed into a close encounter, with both sides near enough to hurl grenades at each other.

The battle all around the Commando troop positions continued throughout the next 24 hours, and there was still no sign of the 8th Army, which was due in only seven hours after the Commandos had landed. It turned out that their forward troops had been delayed by blown bridges and the situation was getting desperate. The Commandos were – as usual – heavily outgunned by their enemy, but they hung on until word reached them that the 78th Division was coming up as the recce force. Their arrival allowed No. 40 to take a break, but they had barely withdrawn from their positions for a breather when they were called back into line. The Germans resumed the attack with such force that the Commandos witnessed the unfortunate sight of Allied soldiers running away from the action, leaving their guns unmanned. 'Pops' Manners pulled out his revolver and threatened to shoot them if they did not get back to the line. What they had hoped would be a brief encounter ended up being a ten-day engagement for the Commandos, and at the end of the period both No. 3 and 41 were so depleted in numbers that they had to put out a call for volunteers from the 8th Army and landing craft crews.

In mid-October, however, both No. 3 and No. 41 were withdrawn to Britain to re-equip and to get new recruits into training. The newly formed No. 43 RM Commando was dispatched to Italy to join No. 2, No.

9 and No. 40 RM, along with a contingent of Belgian and Polish soldiers who formed No. 10 (Inter-Allied) Commando, the elements which now formed the No. 2 Special Service Brigade under Tom Churchill.

No. 43 arrived just after Christmas 1943 and marched straight into something of a maelstrom of activity in Italy and the Adriatic that would engage elements of the Commandos for the remainder of the war. Although slowly being pushed north by the 5th and 8th Armies, the Germans had been running in a constant flow of reinforcements and were stubbornly refusing to accept the Allies' invitation to leave the country. They stood firm and gave as good as they received, often better, in some of the most ferocious fighting the war had seen.

Over the coming months, the famous battle names of Anzio, Cassino and Lake Comacchio would take their place in Allied military history amid appalling carnage during seemingly endless fighting. When two days under fire can seem like an eternity, six months was nothing short of relentless attrition. And so it was in the last months of the Italian campaign: attack after attack, with screaming shells and mangled bodies, scenes unmatched by any action since the First World War. The campaign was faced with great courage on both sides, it must be said, and for which numerous battle honours and a large clutch of Victoria Crosses were to be won by soldiers of the various Allied units engaged.

Those horrendous battles involving upwards of a quarter of a million men have been well recorded and in any event space does not permit me to do them justice by brief descriptions here. Suffice to say, then, that the British Commandos were involved in numerous actions in support of the Allies' long and difficult way forward against an enemy who threw everything into holding them back. Matters really came to a head when the 5th and the 8th Armies reached the Gustav line, which ran from the mouth of the River Garigliano on the west coast to a point 30 miles above Termoli on the east coast. It was the defensive line beyond which the German army commanders had been instructed by Berlin they should not retreat. There, they had to fight to the death. They had 130,000 men and lots of heavy armoury strung out along their defensive positions. Cassino, which straddled the line on the road through to Rome, became the focal point of a six-month-long stalemate in which thousands of casualties were suffered by Allied troops from American, British, Indian, Gurkha and New Zealand forces. Ferocious battles raged day and night and only ended in May to allow the Allies to revive their march north after the massive and controversial Allied air bombardment of the Benedictine monastery at Monte Cassino, part of the fortress around which the advance north had been stalled at a cost of 6,000 men killed, wounded or captured.

Surprise Allied landings at Anzio on 22 January 1944, for which the war planners had great hopes, represented an early attempt to break the deadlock. By midnight, 36,000 men and 3,000 vehicles had been put ashore ahead of the Gustav line, just 30 miles south of Rome. The British and American force included three battalions of American Rangers, the US equivalent of the British Commandos, the latter represented by No. 9 and No. 43 RM Commandos, who were landed from the HMS *Derbyshire*. Although the news reports back in London claimed the landings as a total success, virtually unopposed, the euphoria was short lived.

The Germans rushed to form a semicircle of defensive positions, and hellish battles soon raged. The Anzio force was kept bottled up for almost as long as the Cassino campaign ran its course, but equally it kept a very significant portion of the German forces engaged away from the Gustav line and the crucial Cassino fight. As well as their involvement in the Anzio landings, Brigadier Tom Churchill's Commandos were engaged in the Cassino campaign from the earliest stages, beginning on 29 December 1943, when they had to create a diversion under heavy enemy fire while elements of the advancing Allied troops crossed the River Garigliano. Thereafter, they were moved across the whole Cassino front, from east to west, undertaking a variety of tasks. The fight seemed to be unending, and in many respects it was just what the Allies wanted – keeping 20 divisions of crack German soldiers entertained while back in Britain plans were being made to breach the walls of Fortress Europe. D-Day was coming, but before we return to catch up with developments there we have to rejoin Jack Churchill and No. 2 Commando who had become embroiled in another fulsome arena, in the islands of the Adriatic off the coast of Yugoslavia, which, by the spring of 1944, would have drawn into it three of the Commando units.

Churchill was sent to ensure that Yugoslav Partisans held on to the island of Vis, which was strategically important because it possessed a small airstrip capable of landing and refuelling Spitfires. But it was in the midst of German-held islands and was close to shipping lanes used to supply German garrisons. In many respects the operations in the islands represented a return to basics for the Commandos who took part. They were to stage raids against German shipping and island outposts, but eventually the engagement developed into costly operations against heavily defended German positions whose troops far exceeded their own numbers, carried out on incredibly hostile terrain and under difficult local conditions. They also stepped into the middle of the internal strife that existed between the two main Yugoslav Resistance groups, the right-wing Chetniks under Draza Mihailovich and the Communist Partisans under Josip Broz, later known to the world as Marshal Tito. Originally, they had worked together in the face of the German onslaught into Yugoslavia in 1941, but gradually the enmity between them grew to the point where they were fighting each other. Britain didn't help, first recognising Chetniks and then preferring Tito, who was also naturally favoured by the Russians.

Between them, the two groups managed to produce some of the most effective guerrilla troops seen in any of the occupied countries, hiding in the mountains and operating from the islands off the Dalmatian coast. The Commandos were among a number of British units who were moved in at various times to join the local guerrillas in their attacks on the Germans. Jack Churchill's No. 2 was the first of the Commandos, arriving on Vis on 6 January 1944. They were joined a month later by No. 43 after its release from Anzio. The two Commandos took with them a small flotilla of craft manned by Royal Navy crews – motor launches, MTBs and MGBs – to make their raids on the surrounding islands, alive with German units.

They all took up residence during cold and wet weather in a rugged landscape with mountains rising from the coast, with an equally rugged population who had a somewhat carefree respect for life, believing it an honour to be killed or wounded, and who viewed the incoming troops with great suspicion – which was especially true if there was any contact with local women, and girls who welcomed those attentions faced very severe punishment. The Commandos had not been there long before Jack Churchill was forced to gather his men together for a short talk on the necessity for them to be starved of the joys of sex for a little while longer. 'These girls,' he said, 'will be shot if they are discovered. And so, very likely, will you!' His words were prophetic. Three girls were shot in April for becoming pregnant.

As to the action, the Commandos spent the coming weeks using the island as a base for raiding missions. They would go out night and day like pirates and raid the convoys with the MTBs and MGBs, which could dart out from countless hiding places along the coastlines. They would often pull alongside a supply ship, board it, secure the crew and carry off the cargo back to their island base, leaving the ship scuttled. While these attacks were something of an adventure, the assaults on German bases on other islands were exhausting and difficult. The coastlines where landings were made bore no resemblance to the more familiar beaches of France or even the cliffs of the Channel Islands. They were invariably lined with jagged rocks on which their boats easily foundered. When they landed, their boots might be cut to ribbons as they came ashore.

A large force was being assembled on Vis, with 2,000 Partisan guerrillas under their own leaders, while the Commando strength was bolstered further by the arrival of No. 40 RM in early May. Intelligence reports suggested that the Germans were about to attack Tito's headquarters in Bosnia, and liaison officers forming a top-secret British mission were on their way to visit Tito to discuss plans for joint action between Tito's Partisans and the Allies. The party included Randolph Churchill. On 25 May Winston Churchill sent a 'private and personal' communication to Tito, in which he said: 'Give my love to Randolph should he come within your sphere . . . I wish I could come myself but I am too old and heavy to jump out in a parachute.' In fact, the Germans launched their attack on that very day, and Tito and the British liaison officers, including Randolph, escaped into the mountains. Word reached London on the twenty-sixth that the British mission and Tito were safe and would be secreted out of the country.

The Commandos on Vis, under the overall command of Lieutenant-Colonel Jack Churchill, were ordered to mount a major assault on one of the key German-held islands, hopefully to draw forces from the mainland and thus give Tito the chance to escape. These schemes of drawing enemy fire had always been dubious in the past, and this one was no exception. For one thing, the island chosen for the raid was Brac, which not only contained one of the most inhospitable landscapes among the islands but also had a substantial contingent of Germans and their equipment. Jack Churchill put together a motley army for the raid, consisting of around 1,500 Partisans plus the whole of No. 43 RM Commando, one troop from No. 40 and an artillery battery from the 3rd Field Regiment under the supervision of an expert Royal Artillery fire control officer, Major Turner, a veteran of El Alamein. The men had heavy loads to carry, including their three-inch mortar bombs and Bangalore torpedoes to blow beach defences and other obstructions.

However, after a successful surprise landing that was unopposed, they discovered that the Germans were largely holed up in a pair of eyries in high hillside positions for which Major Turner's big gun was pretty useless. Jack Churchill went to Plan B: flank attacks using the Partisans on one side and the Commandos on the other. The trouble was that when they arrived at the start-line, the Partisans decided that it was getting rather late and put it off to the following day. By then, the swirling sounds of Jack Churchill's bagpipes were already wafting across the hills, signalling his men into battle. He was temporarily silenced by the arrival of a clutch of Spitfires who strafed his chaps instead of the enemy, and as they recovered from the duck-and-dive positions a runner brought the message that the Partisans would not after all be turning out today.

The Commandos pulled back quickly and were once again strafed by the Spitfires. Overnight, Churchill brought out Lieutenant-Colonel

Manners with three more troops of No. 40 Commando from Vis, and the attack began in the face of by now well-prepared German defences. The upshot was that the front-line attack of the British force, consisting of C and D Troops of No. 43, took very heavy machine-gun fire, and because of radio failure in the mountains the men became isolated. Every one of the officers fell in the first attack along with many other ranks. Sergeants Gallon and Pickering, themselves both wounded, continued the assault until they were ordered to withdraw. On that day, No. 43 lost six officers and sixty other ranks killed, wounded or missing. A second attack was led by Jack Churchill himself, again using his 'instrument of war', the bagpipes, with which in his younger days he'd earned money busking when penniless in Paris ('they paid me to go away').

The sound of the pipes was again the signal for the attack by Y Troop of No. 40, and he took them into battle while calling for backup from the rest of No. 40 and No. 43 to follow in behind. 'Y Troop of 40 were magnificent,' Churchill said. 'All in line, shouting, firing, just like on the assault course at Achnacarry.'

They hurled grenades at anything that moved and blasted the scenery with the best firepower they had available. Radio contact was still poor, and only one troop from No. 43 had managed to get up in support. The Partisans, meanwhile, had paid no heed whatsoever to the carefully negotiated order of battle and, according to Churchill, 'had buggered off to God knows where . . . and I never saw them again'. A prolonged silence indicated that the Germans were taking stock, and when they realised that Churchill's force was more or less out on its own they came forth with all guns blazing and put in a strong counterattack. No. 40 Commando, meanwhile, had achieved initial success in shaking up two German positions, but then they, too, were hit by flanked counterattacks and took heavy casualties. Among them was Lieutenant-Colonel 'Pops' Manners, one of the great stalwarts of the Commandos, who was so severely wounded he had to be left for the Germans to pick him up. Jack Churchill himself was then captured, complete with claymore and bagpipes.

The British units suffered more than 81 casualties, killed, wounded or captured in the attacks, and retreated back to Vis. German medical units immediately came out for the wounded of both sides. 'Pops' Manners died under treatment and word was later sent out that the German medics had done all they possibly could to save him. The British prisoners were allowed to dig a mass grave to bury their dead. They all stood around in silent prayer as Jack Churchill played a lament on his pipes: 'Flowers of the Forest'.

Not among those in the pit was Knocker White, of Y Troop, No. 40, although all his mates in the troop believed he had been killed. Robin Neillands tells the story[1] that Knocker took a bullet wound to the neck, and one of his closest pals later reported that he was dead and wrote to his wife to say 'Knocker died in my arms'. But, in fact, Knocker wasn't dead; although he'd lost a lot of blood, he came round after his unit had departed to discover a couple of Germans prodding him with their bayonets. He said a few choice words to them, asking if they would kindly stop doing that and take him to the nearest medic before he was down to his last drop of blood. Knocker duly survived, and months later his wife received the good news, bad news letter from the War Office: 'Your husband's alive and would you now please return the widow's pension we have been paying you for the last six months.' Mrs White apparently responded in true Commando style.

Though the whole operation went badly awry, largely through the failure of communications and the unreliable nature of the temperamental Partisans, it was successful inasmuch as it had the required result. Tito was rescued from Bosnia and brought to the safe haven of Vis, where he had a number of visitors in that month of June. Among them was the overall commander of the Commando group, Major-General R. G. Sturges, who brought news of the D-Day landings and the part that the Commandos had played in it, and everybody on Vis was wishing they were there and not here.

[1] *By Sea and Land,* Weidenfeld and Nicolson, 1987, p. 120.

In the new wars that followed peace, the Commandos, by then under the sole banner of the Royal Marines, were in constant action for the coming two decades in post-colonial trouble spots. (Above) Back to the jungle of the Far East against the Communist terrorists in Malaya (here in 1950) and (below) members of 45 Commando discover a hideout of EOKA terrorists in Cyprus. (*Royal Marines Museum*)

An artist's impression of the Suez fiasco, 1956: RM Commandos were at the forefront among the unfortunate British troops selected for the invasion of Egypt. It was also the first time the Commandos used helicopters to deliver their troops, largely because of a shortage of landing craft and they flew into a hail of gunfire. (*Royal Marines Museum*)

A patrol from X Troop, 45 RM Commando advances through the streets of Port Said during the invasion of the Suez Canal zone, before the United Nations ordered Britain and France to ceasefire. (*Royal Marines Museum*)

Commandos continued to undertake spearhead duties in post-colonial wars throughout the 1960s. A patrol from 45 Commando makes its way across the unforgiving territory of the Radfan, above Aden, where the unit served continuously for almost seven years. (*Royal Marines Museum*)

All Commando units were at various times drawn into the Borneo confrontation in the 1960s, fighting in dense jungle conditions. Here, a unit from 42 Commando man a rifle training post for border guards in the then British territory of Sarawak. (*Royal Marines Museum*)

The Gulf – first time round: in July 1961 elements of 42 and 45 Commando landed in Kuwait in the face of a threatened invasion by the new military rulers of Iraq. The attack never materialised but baking heat in open desert bore down relentlessly, especially on positions like this one occupied by a gunner from 42 Commando. (*Royal Marines Museum*)

As the Commandos were returned from active service in the Middle East, tensions were rising in Northern Ireland . . . and they were among the first British troops to be deployed on the streets of Belfast, pictured here in the Westbrooks area. Since 1969, not a year has passed without the presence of Royal Marines on rotational duty there. (*Royal Marines Museum*)

The Falklands War drew all three RM Commandos into the Task Force of British troops: ıere, on the windswept, boggy landscape, a typical scene of digging in for a defensive)osition. (*Royal Marines Museum*)

Back into Northern Ireland . . . and now with regular helicopter patrols, as in this picture nside the cabin of the aircraft and what had become routine for all of the three Commando units over three decades of the Troubles.

Commando training is renowned as among the toughest for any military unit anywhere in the world, with many of the techniques developed from original training centres set up in the Second World War. There are five separate training wings run from the Command Training Centre at Lympstone, Devon, covering courses lasting 30 weeks and culminating in a massive test of physical and mental endurance at the infamous Commando assault course. (*Royal Marines Museum*)

Chapter Eleven

D-Day: Death and Destruction

Preparations for the invasion of Europe, the largest amphibious landing of troops ever mounted, swung into top gear when Generals Eisenhower and Montgomery arrived back in Britain in January 1944. The initial invasion plans called for ten Commandos to back up the spearhead landings of the parachute and air-landed troops who were to go in first. This was proved to be an impossible demand because at the time the Commandos were already heavily committed elsewhere. It will be recalled that there were four Special Service Brigades now consisting of sixteen Commandos. As we have seen in the previous chapter, 2SSB[1] was still heavily engaged in Italy and off the coast of Yugoslavia. 3SSB[2] was en route to the Far East, ultimately to join Mountbatten's South-East Asia Command. This left Nos. 1 and 4 Special Service Brigades available for D-Day – eight Commandos in all, three army and five from the Royal Marines: 1SSB under the command of Brigadier (Lord) Lovat consisted of Nos. 3, 4 and 6 army Commandos and No. 45 (RM) Commando. No. 4 SSB, commanded by Brigadier B. W. 'Jumbo' Leicester, was an entirely Royal Marine Brigade of Nos. 41, 46, 47 and 48 (RM) Commandos. Of the marine Commandos, only No. 41 had seen action previously under the Commando banner. The rest were fresh out of training, which for many had simply been a crash course in the basics of Commando techniques, although most had, of course, served in regular marine units. This alone caused consternation among the more experienced men of the Commando. To some extent their fears would be well founded, although the marines' courage was never doubted.

Typically, the final rush for the Allied invasion of Hitler's Fortress

[1] Consisting of Nos. 2, 9, 40 and 43 (RM).
[2] Consisting of Nos. 1, 5, 42 and 44 (RM).

Europe, three years in the planning, left no space for finesse as Operation Overlord edged towards the brink of reality. Some of the small print would be overlooked in the logistical nightmare of manoeuvring hundreds of ships and aircraft, thousands of tonnes of assorted military hardware, millions of rounds of ammunition and two million troops into place for the hugely complex operations now emerging. The date, the time and the place were the subject of much speculation by all, and especially the Germans. Rommel himself, who was to command the German opposition, provided the title for the eventual book and the movie when he said, 'for the Allies, as well as the Germans, it will be the longest day'. Indeed it was: 156,000 American, British and Canadian troops to be landed on the coast of north-western France within the first 24 hours, the first wave and those most at risk from both the German shore batteries and the 4 million mines Rommel had ordered to be laid in the months beforehand.

It would soon become clear that airborne forces and parachute troops would spearhead the Normandy invasion with Commandos providing the backup to assaults on gun batteries and key German positions to help give the advancing main force a chance. By late spring, a new British airborne division, the 6th, was fully operational under the command of Major-General Richard Gale, one of the pioneers of the parachute brigades, with Brigadier James Hill's 3rd Parachute Brigade as the core unit. Hill's three battalions were to be joined by a Canadian parachute battalion, the 5th Parachute Brigade, and the 6th Air-landing Brigade. The 6th Airborne Division was to go in ahead of the invasion force, dropping around Ranville, north-east of Caen, to cover the landings of the British 2nd Army coming ashore on their designated beaches, codenamed Gold, June and Sword.

The US 82nd and 101st Airborne Divisions were to perform similar operations for the US 1st Army, landing at their beaches of Omaha and Utah. The area to be taken by the British 6th included high ground on the eastern sector around the River Orne and the Caen Canal. From midnight on 5–6 June, paratroops and glider-borne units began landing at key points. Throughout the night, 230 RAF bombers began pounding German positions, and at daybreak 1,300 heavy bombers of the US 8th Air Force took over the air attack with escorts from Mustangs, Lightnings and Thunderbolts. Shortly before dawn, thousands of ships assembled around the southern ports of Britain, packed to the gunwales with fighting troops, transport and equipment, set sail.

The 'must-achieve' targets for the paras and airborne forces on the British side of the invasion plans therefore included the infamous Merville battery, which was itself capable of knocking out thousands of troops as they landed, the high ground overlooking the route forward for the advancing British troops, and the bridges across the canal and river to halt an immediate counterattack by enemy forces while the beach landings were proceeding. For this purpose, Lord Lovat's 1st Special Service Brigade would provide additional firepower for the 6th Division after landing by sea 90 minutes after the first troops had gone ashore.

Lovat's group, with 146 officers and 2,456 men, was to land on Sword Beach beside the River Orne, on the left flank of the Allied armies. They had orders to capture Ouistreham and then push inland to support the Airborne Division in seizing and holding the crucial bridge over the Caen Canal. This was Pegasus Bridge, famed as one of the Parachute Regiment's proud battle honours.

No. 4 Special Service Brigade, meanwhile, would head west to attack coastal batteries and small resorts such as St-Aubin, Lion-sur-Mer, Luc-sur-Mer and Angrune-sur-Mer before moving in a lateral deployment to link with elements of the British and Canadian forces. No. 47 (RM) Commando had a particular mission to attack the fishing port of Port-en-Bessin, and No. 46 was kept offshore as a floating reserve and eventually landed on D + 1 (7 June).

In addition to the Commandos, the Nos. 1, 5 and 6 units of COPP had, in the months ahead of the invasion, carried out beach surveys and mapping of hazards and submerged obstacles; they would also mark the beaches on the huge expanse of French coastline designated for the landings. Theirs was a most vital operation and well deserved the many medals distributed among the relatively small group of men who contributed. On the day, and for the remainder of the landings, RN Commandos were also out in force, fielding eight units and one in reserve for the equally vital role of beach-masters and traffic managers.

Gradually, men and machines were moved into position for the off. Henry Cosgrove was with 45 (RM) Commando, in Lovat's brigade, and at the end of May, after training in Scotland, they were all moved south:

> We moved to a place near Southampton, a wired compound; nobody was allowed out. Security was intense. However, in one big marquee there were lots of big boards with overlapping aerial photographs of the French coast. Didn't tell you where exactly. You virtually had a whole area in pictures. People were continually going in to look at this. This is some place where you're going to go, but nobody's going to tell you where. More and more photographs were added. And then things started to happen. We were issued with special 24-hour rations; we weren't allowed to touch them. There was . . . a tin, like a sardine tin. It was a solid block of chocolate. It was as hard as iron. The only way you could break it

> up was to get at it with your bayonet. The rations were in a small box about eight inches by five by two deep. In it were cubes, little white cubes about the size of an Oxo cube with black specks. That was tea, sugar and milk. One of them would make a pint of tea. There were cubes, slightly larger, which was chicken soup. You had little cookers with cubes of solidified methylated spirits. They were round, about four inches high, with a floor in the middle where you put your billycan on the top and whatever you were going to make, tea or soup, in it. They were landing rations. Then we had our kit checked, time and again. Everyone had a clean pair of socks, a clean pair of underpants, a vest to match and a clean shirt. That was your clothing. Then . . . your rucksack: in the bottom I had 4 two-inch mortar bombs, explosives and 200 rounds for the Tommy-gun. I also had 2 smoke bombs. They were dangerous. They'd got phosphorus in them. I was worried about getting a bullet anywhere near them. That was a tidy weight; about 80 pounds in the rucksack. Then I had 200 rounds of ammunition; every man in the unit including the colonel carried 200 rounds of rifle ammunition in a bandoleer. Everybody's boots were checked, and we had another look at the aerial photographs.
>
> We moved from Southampton to a beautiful little place, Hamble, and embarked in five lots. I've never seen so many ships in my life, an amazing sight. It appeared that the navy had made a roadway going across the Channel. You seemed to be travelling through a line of ships each side of you, as if going along a marked-out road. Once we left the shore and were on our way, we were told what was going to happen. We were told our job was to get to the Orne bridges as quickly as we could to relieve the paratroopers.

It was that journey across the Channel that struck William Spearman, a veteran campaigner of numerous Commando raids, including – it will be recalled – Dieppe with No. 4 Commando. Most of those in the massive convoy will remember the crossing in rough weather, and for many in the various types of landing craft it was an appalling journey, being tossed and buffeted by the sea and the rollers created by the sheer mass of ships. Hundreds of men, packed like sardines into their craft, were violently seasick, which merely added to the overall discomfort and fear. Spearman said:

> I think going across the Channel, the enormity of everything began to become apparent. No matter which way you looked there were ships of all shapes and sizes as far as the eye could see. And as we

neared the French coast, on our left, not too far away from us, we saw, suddenly, one ship explode into the air and sink, all in a few minutes. So we assumed this must have hit some big mine which accentuated the knowledge that they had mined the coast. We were led by minesweepers, but it was very hard for them to remove every mine. So a number of ships were sunk by mines before they even got there. We had strict instructions of what we had to do. We were on a left flank, the complete east flank of the whole of the landing. Our first job was to knock out a gun battery at Ouistreham, which was right on the east side of the landing. And then after that we were to take up positions in the Hauger area a few miles inland and defend that against all-comers. And if our boat had sunk, I really don't know who would have taken over that job because there were no reserves for us. It was a very thin green line. Who would have knocked out the gun battery? Who would have defended that left flank?

Fortunately, our boat travelled safely until the time came to get into our ALCs. Then our attention was again drawn to the magnitude of the whole operation, left and right, everywhere, all these bloody great ships and landing craft. And then our bombers going over all the time, to do the softening up. But when we landed you'd never believe there'd been any softening up. Troops that landed in front of us, 3rd Brigade, East Lancashire Regiment, I think, and Canadians, were in severe trouble. I often wondered afterwards if they'd been sent as gun-fodder to distract their attention from us, because as we came ashore there were bodies everywhere. We were lucky to even land. Some of the boats as they went in were blown out of the water long before they reached the beach. The mother ships got blown up, the transporters got blown up. The noise was just indescribable. Luck was with us and we landed on the beach, struggling forward under the tremendous weight on our backs; half-hundredweight or more. We carried our own weapons, flame-throwers, shells . . . the idea was we had to have enough to support ourselves in case we didn't get reinforcements or we didn't get replenished. And once you go down with that pack on your back, you can't get up again. We dashed ashore and we were all shocked by the number of bodies, dead bodies, living bodies, and all the blood in the water giving the appearance they were drowning in their own blood for the want of moving. The whole place was littered [with bodies]. There were great, monstrous fortifications on the beach, like tremendous cubes of crisscrossed steel girders to stop gliders landing, to stop ships coming in. There were girders and poles penetrating into the beach, sticking up to stop boats coming in. The whole beach area was covered by flame-throwers, and dominating the whole landing areas were these huge pillboxes. You just couldn't get out of range of them. The fortifications were excellent. For our chaps coming ashore it was a desperate situation.

And it wasn't our job to stop and put them out of action. We had to get to the gun battery. We had one object in mind, to get off the bloody beach. It was quite terrifying, the enormity of everything, and having seen all the bodies on the beach doubled our determination. No matter what happened, we had to get off the beach. Once you get off the beach, you're out of the fire, as it were. But it's a very hard lesson to learn, and people bloody won't learn it.

Why did they put so many untrained troops on the beach first? Any one of us could tell you they wouldn't get off. They'd be so transfixed with fright they couldn't get off. We, of course, were transfixed with fright, but we had the certain knowledge that you either stopped and died or you made a dash for it. Some of our chaps did try to put some of the things out of action on the way. I remember one Lieutenant Carr, a terribly thin man but a very good officer. He deliberately went up to one pillbox and threw a couple of grenades in, and I thought it was a very brave deed, just to hesitate in order to do that.

Lovat's Commandos set off through the hail of fire with every kind of metal whistling and exploding around their ears, heading towards their designated targets. They had to rely on their maps and their memory of the reconnaissance photographs that they had viewed before setting off. It was really a case of establishing a general direction and following the noise, as Henry Cosgrove remembered only too well:

The aerial photographs were an immense help now because you're in a strange land, you don't know anywhere. On these aerial photographs there was a piece of slightly higher ground with a wood. We knew we'd got to make for there, that was our assembly area, which was about two miles inland. Once we were organised there, we set off for the Orne bridges. These [airborne] lads had set up on one end of the bridge. When we got to the bridges there were a lot of dead Germans around and some of our paratroopers. There were snipers around, firing at us, but nothing serious. And then the brigadier set off over the bridge as if he was out for a stroll on his country estate, with bloody bagpipes in front.

It was Lord Lovat [with Piper Bill Millin, who led a number of

Lovat's attacks]. We followed him over the bridge and then the problems really started. We were to hold the high ground up round and check if the paratroops had secured the Merville battery, or if we had to lend them a hand. But in fact they *had*, partially, although Germans were in great strength all around there, and we started losing a lot of men. We moved to higher ground and dug in because we were being mortared. We were in a circle near a small village under heavy fire. Then, to crown it all, we lost our radios. The first signaller was shot and then we lost the second signaller, wounded. Both sets were out of action, and we couldn't make contact with brigade headquarters. By this time we were totally cut off. Fortunately, we managed to reach an artillery officer at a forward OP who was laying down fire for us from the ships if we needed it, and we managed to relay our position via him, to the ship and back to brigade headquarters. They told us to come out at night – but that was a long time away and we were running out of ammunition. In fact, we were down to sharing it out among ourselves. Our CO, Lieutenant-Colonel Ries, had been wounded and Major Gray – we called him Dolly – was running things. He decided we had to try and break out and get back to brigade headquarters at Le Plein. We didn't do too badly for about a mile, and then two-inch mortars were becoming very accurate; we were mortaring very well with what we'd got then and managed to get through, although we lost a few men in the process. When we rejoined the brigade, we'd been out 36 hours without sleep.

Bernard Davies, with No. 4 Commando, and his group were late in arriving. Davies's company did not get ashore until late in the evening. Their landing craft was damaged as they were coming in and was left floundering off the coast for hours on end amid the mayhem, until the activity died down somewhat and they were able to get a tow inshore from another craft. By then, the beach was an incredible sight of beached and half-sunken landing craft, tanks that had become bogged down in the sands, wounded men still abounding, as were the bodies of the dead. Davies's company headed for the action and ran straight into trouble:

When we arrived, 6th Airborne Division were in a very dodgy situation. Our object was to grab the high ground east of the landing strips which, if they'd got some guns along the top, could have commanded seven miles of beach. They were almost overrun. My unit, C Company, suffered heavily as we took up our positions. We lost 80 per cent casualties, lost all of our officers and the

> sergeant-major, Peter King, was promoted to lieutenant to take over; he was awarded the DCM. The Germans had thrown everything at us, and we were armed with those blasted Bren guns which only fired about 400 rounds a minute. The Germans had one that fired 1,500 rounds a minute – 25 rounds a second. We were, as usual, very inadequately armed. It was close fighting, 30 or 40 yards, heavy casualties, but we held our ground until we were pulled out when a marines Commando came up and we were able to take a rest until they moved us up to Breville, in the front line, still surrounded. The Germans held most of the ground all the way into Caen. We were to stay there for two weeks, battling away and eventually moving forward. We took quite a few casualties from night snipers. I learned more in a week in Normandy than I'd learned in the whole of my previous four years. I was amazed at some of the things the Germans did. In the country lanes, there were high banks at the side of the road which had walkways cut through, except they were cut at 45-degree angles instead of straight, so you could never see them coming until it was too late. They had machine-gunners in the hedgerows during the day; they fired down the hedgerows from positions where it was impossible to get sight of them, and then at night they pulled them out and put them into the fields so our night patrols were sitting ducks. They could hear us coming and cut us up. The RAF often saved the day for us against forces that were far better armed. We had the men and the organisation – 6th Airborne Division with Commandos stuck in the middle of them – could be not bettered anywhere. We were badly let down on firepower.

That point had not been lost as the commanders of 1SSB assessed their situation after four days of hard fighting. Lovat called a conference of his commanding officers at his brigade HQ. Preliminary figures showed that the brigade had lost 270 men so far; it was an incomplete summary and proved to be an underestimate. Lovat could not contain his anger when one of his commanders, John Durnford-Slater, came up bearing words of congratulations from army commanders. Lovat stormed:

> If the Germans repeat the same performance tomorrow, General Dempsey can expect a butcher's bill that spells curtains for the brigade, and the high ground will fall. And where the hell are the reinforcements to cope with this wholesale destruction? They're supposed to have a depot and a training centre full of soldiers and they supply fuck all. Twenty-five lucky lads as replacements for

> four days' fighting – less than half of one troop – committed piecemeal to the battle the moment they arrived. Killed or missing, poor devils, without knowing what hit them. Fifty new men were needed before we cleared the beach. We lost more than this handful, sunk at sea. Who is responsible for this balls-up?

No one could provide the answer. On Monday morning, 12 June, the divisional commander issued orders for the capture of a strongly held area of Breville Wood where the Black Watch had taken serious casualties the previous afternoon. Henry Cosgrove's unit was among those who went in, though still short of men. For a while it went quieter. Then, said Cosgrove, as they prepared for the Breville assault, General Gale, of the 6th Airborne, sent out probing patrols 'by the minute, finding out what was about' before the Airborne Division went in for the frontal attack with the Commandos going in for a flanking attack to capture a German headquarters:

> All hell broke around there. Casualties were very heavy. The Airborne lads were very badly hit and then they started belting us. They had mobile guns and were mortaring very accurately. We were taking quite a lot of casualties and getting short of men, some killed, some wounded and others missing. By now we'd lost over a quarter of our strength and these casualties kept on coming through. Bloody hell! We were getting plastered by the mortars. One of the doctors arranged a convoy of Jeeps to shift the wounded. The Jeeps were flying to and fro, and there was mortaring going on the whole time: bloody brave men, they were, sitting up there. It's not so bad when you've got a hole in the ground and you've got a bit of cover, but out in the open, moving the wounded . . . It was quite a night, I can tell you.

Earlier that same day Lieutenant-Colonel Derek Mills-Roberts, commanding No. 6, had been sidelined for treatment for a leg wound which was developing gas gangrene. He was given an intra-muscular injection and ordered to remain in a prostrate position for some hours at the farm they were using as a base. His unit headed off into the Breville attack and he was left lying down, drowsy from the drugs he had been given. A few hours later, and still groggy, Mills-Roberts was woken by a corporal bearing a message from Breville. Lord Lovat had been wounded in a violent German counteroffensive and he was to get over there immediately. The brigadier wanted to see him. Mills-Roberts roused himself, feeling sick and still drowsy, and dashed to the scene and found that

Lovat had been removed to some old stables:

> It was almost dark and the ghastly scene was lit up by the farm buildings, which were on fire. Lovat was a frightful mess; a large shell fragment had cut deeply into his back and side. Peter Tasker, No. 6 Commando's medical officer, was giving him a blood transfusion. He was very calm. 'Take over the brigade,' he said, 'and whatever happens – not a foot back.' He repeated this several times. And then, 'Get me a priest,' he said.
>
> The parachute battalion making the attack on Breville had been badly hit by mortars and *Nebelwerfers* – petrol bombs from this last insidious weapon had burst, causing desperate injuries and burns among the parachutists. Their colonel was mortally wounded. Other wounded men were on fire, and we put out the flames by rolling blankets round them. Further out lay more wounded. The RAMC were doing magnificent work, while Sergeant-Major Woodcock found every available Jeep to get stretcher cases to safety. I no longer commanded No. 6 Commando, and, as acting brigadier, had to get on with the wider battle which was reaching round the whole area. I went back into the farmyard, where every available foot of space was covered with stretcher cases, but the Jeeps were making a good clearance. If more shells had fallen into the crowded farmyard, the slaughter would have been terrible.

Lord Lovat was among those wounded who were eventually carried back to field hospitals, and he survived his very severe wounds.

Brigadier Jumbo Leicester, meanwhile, was still going strong. His No. 4 Special Service Brigade had landed with rather less togetherness, as it were, than the units of 1SSB and were sent off in various directions, suffering mixed fortunes. Most had the same kind of beach landing experiences described above, and a good many of the troops of Leicester's RM Commandos never made it inland. Because of their more diverse directions, space does not permit more than a cursory summary of their activities in the following seven or eight weeks. No. 41 (RM) Commando, it will be recalled, was the only one to have seen action so far in the war – the rest all having been raised in the months preceding the D-Day invasion. It must be said, however, that most had served in the marine battalions and, indeed, the Royal Marines as a whole fielded 17,000 men for the Normandy landings in a variety of roles ranging from shipboard protection to assault forces.

Three of Leicester's Commandos had progressive tasks to complete and move on. No. 41 was first designated to attack German positions at

Lion-sur-Mer from their landing on Sword beach. But their plans went awry from the outset. They were dropped ashore in the wrong place, almost a quarter of a mile from their given landing area. The dash across the beaches was to be performed under a veritable hail of shells and mortar fire, at a point already strewn with the debris of bad landings: blazing vehicles, bodies and stuck-firm tanks. Once ashore, No. 41 was divided in two, one to attack a German strongpoint on the approach to Lion, the other to attack a stoutly defended château used as a German planning base. Both tasks resulted in a hefty response from the Germans in residence and were not completed for almost 24 hours. The Commando came back together for their next objective, which presented even greater opposition. They were to join No. 48 in a raid on the vital Douvres radar station, which was surrounded by massive fortifications and minefields and which the Germans believed was pretty well impregnable. It was certainly no knock-over, and after 12 days an invitation to the Germans to come out with their hands up was adamantly rejected. The Commandos sent for flail tanks to clear mines and then launched a full-scale assault with a troop of tanks and naval bombardment. This time, the Germans gave up. They then moved on to Sallenelles, across the Orne, where General Gale used them for patrolling, a highly dangerous business, as Bernard Davies has already described.

No. 48 Commando had landed on Juno beach with the initial objective of an assault on St-Aubin, following in behind the Canadian North Shore Regiment. But they had a bad time landing. The Commando was formed only three months before D-Day, and any veteran of Commando raids would put his hand on his heart to say that was simply not enough time to get a man fully fit and trained for the task at hand. This was regardless of whether he was already a serving marine or not; in fact, they would say it was a sheer impossibility and would certainly present situations in which the risks to the men would be unacceptably high.

No. 48 discovered some of them immediately they came ashore bearing the huge burden of their backpacks. The wet landing experienced by many of them meant the weight they were carrying virtually doubled; at the very least it would be increased to not much short of a hundredweight. This, coupled with the inexperience of the Canadian units, made it a doubly dubious situation. But in fairness to No. 48, most of the problems were by no means the fault of the men. For starters, they were in totally different landing craft from those in which they had been trained; they literally had to manhandle gangplanks over the bow of the craft and they were all getting shot doing it. The spot they landed on was also wrong; it was right under the noses of Germans inside a heavily fortified concrete strongpoint who let loose immediately. The Canadians hadn't secured the beach and chaos was developing rapidly. Men were dropping as they ran forward amid the tanks and other vehicles struggling to get off the beaches. One of No. 48's officers actually ran back to try to stop a Canadian tank that was attempting to get moving, seemingly oblivious to the fact that it was running over wounded men in the process. He banged on the turret and when the tank still didn't stop he disabled it by chucking an anti-tank grenade into the mechanism. What was left of No. 48 were assembled under the harbour wall and made off towards their target, passing through the Canadians to St-Aubin and on to another big scrap at Langrune a couple of miles east along the coast. For all their losses, No. 48 put up a spirited fight and eventually captured the strongpoint there and saved a lot of lives among those coming through.

No. 46 had arrived with special training for a clifftop assault on a German battery, but in the event it was cancelled and they were given a substitute objective, joining a French-Canadian unit in clearing enemy positions along the tidal River Mue. It turned out to be a far more onerous project than the one for which they'd been trained. The enemy troops were holed up in the villages of La Hamel and Rots and, unbeknown to the Commandos, they were Waffen SS of the 12th Panzer Division, crack, dedicated young troops who rose out of the Hitler Youth. It was a bitter struggle with street fighting and close-quarter combat. In the process, No. 46 lost more than 60 men, killed, wounded or missing. But they put the Germans to flight and captured 47 prisoners.

And so it went on, the advance moving ever outward and the Commandos and the paras still fighting far longer than they had anticipated. Some were still there in October, and No. 47, who came in after the first night as a floating reserve, was expected to be withdrawn along with the rest quite swiftly. In fact, they did not return to England until 14 months later, as they got caught up in the front line of the advance, having told their wives they expected to be home well before Christmas. Trouble was, they didn't say which Christmas.

Many did not make it at all. Both 1SSB and 4SSB suffered proportionately the same number of casualties in the D-Day campaign: 39 officers and 371 other ranks killed, 114 officers and 1,324 other ranks wounded, and 7 officers and 162 other ranks missing. This represented slightly less than half the overall establishment of the two brigades and was a devastating total for them, although a minuscule total compared with the overall Allied losses of 5,500 killed, 22,000 wounded and 12,000 missing in just 15 days after the D-Day landings. Those who made it home were in a poor state, as Henry Cosgrove recalls:

We had kept on advancing forward all the time and the Germans were on the brink of throwing in the towel in France. It was the end of August and we hadn't been out of the line, virtually, since we arrived. But that was it. Our job done. They took us to a rest camp on the coast and then we were homeward bound. To be honest, I'm not sure whether it was Portsmouth or Southampton, but there was a band there. We came down the gangway and we were filthy, and I mean filthy. We'd been three months . . . our uniforms were rags, most of us had fleas, some had lice. We were in a bad way. We'd washed in puddles when we had a chance and that was it. As we fell in on a railway station on the jetty by this train, some silly blighter let off a firework. One minute there were all these men standing there; the next minute there wasn't a soul in sight, they were under the train. We were literally bomb-happy then; if you'd gone up to any one of us then and shouted 'Boo!' we'd have jumped in the air and then turned around and belted you one. Nerves were ragged. We'd been bombed and shelled sometimes for days on end. We went to Petworth camp, where we were ordered to strip off and all our clothing was burned and our personal belongings fumigated. Then there were showers, a medical inspection, many heads were shaved and painted with gentian violet. When that was finished, we came to a big marquee, and as you walked in you got a pair of underpants, vest, a pair of trousers, jacket, socks and boots. Then, outside, you all stood on this field swapping round until you found things that fitted, to get some semblance of a uniform before going home on leave. When we returned, we were made up to strength which took quite a while because we had lost some good men . . . but they'd already got us lined up for another job.

Chapter Twelve

Last-chance Saloons

D-Day was only the beginning, of course. Central and eastern Europe still had to be tackled, and in the Far East things were not going at all well. Lord Mountbatten's arrival to take over the south-east Asia debacle was supposed to change all that; at least, he had the ambition and the drive. He was, however, greeted with cynicism and suspicion. There were so many problems, so many imponderables and so many conflicting views and differences that it was difficult to know exactly where to start. On his arrival in October 1943 he had discovered that priorities ranged through what he described as the three Ms – monsoon, malaria and the morale. The monsoon was always appalling, malaria endemic and morale of British troops was not just on the floor – it had sunk into the paddies and had been trampled underfoot by the Japanese. Memories of the dramatic British withdrawal from Rangoon and being chased out of Burma by the Japanese at a cost of 13,463 casualties still weighed heavily.

Mountbatten's ambitions were not drowned by what he discovered. He had been warned what to expect by former inmates prior to his departure. He wrote gushingly of his thrill of crossing into India to his base at Delhi, because it 'has fallen to me to be the visible symbol of the British Empire's intention to return the attack to Asia'. That was a tall order, and by the time he had finished trying to meet it he himself realised that there was no hope for the continuation of any semblance of Empire rule in that region. The expatriate communities and the colonial staff had simply been drinking in the last-chance saloon, oblivious to the threat of Communism and nationalism which would sweep across the Far East in the wake of the great march across China by Mao Tse-tung. But those problems were to come much later. In the early days the thrill ruled his head, and he wanted to get back to Rangoon for the sake of the Empire.

It was a challenging task, given that all Britain's possessions in the region, with the exception of India and Ceylon, were in enemy hands. Worse than that, he had inherited the most under-equipped and seriously neglected military units in the entire British military establishment. It was, indeed, 'the forgotten army', as Field Marshal William Slim described it when, perhaps understandably, all Britain's resources had been piled into the D-Day effort.

Mountbatten was to oversee a combined military staff of 15 generals and a couple of admirals senior in service to himself, most of whom had differing views on the troubles facing them. They were all based half a continent away from the front lines, while the officers and men of the various Allied divisions amassed on the Indian borders in bug-infested camps from which they were expected to trudge into the jungles and swamps of the Arakan and Burma. They were all fully aware that the staff who were administering their needs were accommodated in luxurious surroundings and palaces in Delhi, where life went on as if not a care existed in the world, and that war was a far-off discomfort.

One of Mountbatten's first tasks was to stop social callers to the staff offices and put a virtual ban on their attendance at the sporting and social events that the staff had until now been attending with alacrity. Then, pursuing his three Ms, he attempted to tackle the problem of malaria. He was visibly shaken to discover that 84 per cent of British troops under command were suffering to varying degrees from malaria, some very seriously ill. He was appalled by the hospital facilities. He found them 'disgusting and inadequate', lacking in hygiene and racked with slovenliness. He sent an urgent dispatch to London requesting 700 nurses immediately and demanded that the Indian administration improve the facilities.

Not least among those with whom Mountbatten would come into immediate contact and conflict were generals who were opposed to his own ideas of extending the fight against the Japanese with immediate effect. The Americans' view that the British had been running a policy of inactivity in Burma since their retreat was not far from the truth, and was outlined in a top-secret memo to Washington from General Albert Wedemeyer, who was overseeing the Americans' involvement with the Chinese army: 'The entire attitude of the British has been pathetic . . . Mountbatten's attitude has been fine, aggressive and enthusiastic but between him and the troops are commanders who emasculate his directives and who are so anxious to prove that they have been right the past two years that they intentionally do not follow the orders of the Supreme Commander.'

It was because of the general level of inactivity that the Americans favoured operations planned by Orde Wingate, the somewhat eccentric former intelligence officer in Palestine who had theories on long-range penetration into the jungles of columns of troops resupplied by air. The local generals despised both his unconventional ideas and his manner. His Commandos were called Chindits, after the Burmese word *chinthe*, meaning winged stone lion. They were drawn from elements of the King's Liverpool Regiment, the 2nd Battalion of the Burma Rifles and the 3rd Battalion of the 2nd Gurkha Rifles. The original force had in all around 3,000 men and began operations in the early summer of 1943. Wingate's first major raid was to take a huge column of troops and weapons, hauled by mules and elephants, to blow up the Mandalay railway. It was a disaster. They got bogged down in the jungle for almost 12 weeks, marched hundreds of miles, often going round in circles, and Wingate lost a third of his force to Japanese counterattacks and malaria. The newspapers in England heralded it as a masterful raid behind Japanese lines. One writer, much later, was more pointed in his comments: 'There was a psychopath running these mad expeditions behind enemy lines: Orde Wingate.' But the British army was fighting a fanatical enemy whose tactics were beyond the bounds of anything previously experienced, as General Sir Walter Walker, then a lieutenant-colonel in the Burma campaign, explained:

> They [the Japanese] were first-class jungle fighters, which is why we were driven out of Burma in the first place. We had to learn their techniques, and then stand and fight. And when we did, they resorted to other methods, such as tying TNT to their stomachs and throwing themselves in front of tanks to blow them up. People tend to forget, if they ever knew, that there was no fighter like a Jap fighter. They were also incredibly cruel and ruthless. I've seen my soldiers' bodies after they'd done bayonet practice on them . . .

The generals in Mountbatten's command were not at all convinced by Wingate's methods. Slim himself, though a supporter of Mountbatten, thought in certain conditions his force would be useful, but in Burma in 1944 the Chindits were more trouble than they were worth. When Mountbatten arrived, a second Chindit expedition was being planned, and initially he gave Wingate his full support, which was not unexpected given his liking for larger-than-life ideas and characters. As Wingate's demands became more aggressive, however, Mountbatten began to have second thoughts. Even so, by 10 March 1944, 9,000 men and 1,100 animals of varying sizes were airlifted far behind Japanese lines under Operation Thursday. It was a complex, meandering proposal involving six brigades to be deployed by air and on foot into jungle locations. The object, according to Wingate, was to insert a substantial force into the guts of the enemy. Mountbatten liked his turn of phrase but was not at all sure about the rest of it, and his military advisers were totally opposed. Wingate was killed on 24 March 1944 in an air crash and was succeeded by former Gurkha officer Brigadier Joe Lentaigne. The operation continued, each brigade forming a single-file column on their jungle marches, each with its own objectives. Ex-Commando Mike Calvert was among the commanders with 77th Brigade. During the coming weeks, there were some harsh battles and running skirmishes. The Chindits were now operating on a far better basis than previously, although casualties still tended to be high against the suicidal attacks of the Japanese. Many bitter battles were fought and won and a large number of medals awarded, including two Victoria Crosses, one posthumously to members of the Gurkha contingent[1] whose like-for-like fighting against the Japanese became a renowned feature of the south-east Asia campaign.

Calvert's brigade had been in continuous action for almost six months when the pugnacious American general, Joe Stilwell, Mountbatten's Deputy Supreme Commander, called for assistance as his Chinese-based force attempted to push towards Japanese positions at Mogaung, 160 miles away from where Calvert's troops were then fighting. The only way to reach them was to walk, and it was monsoon season. Calvert's brigade was already in serious trouble from malaria and dysentery, and the troops were dog-tired. But Stilwell had demanded their presence and he had clout. They arrived at their map reference on 8 June 1944 and began a battle that was to last 16 long days, at the end of which the original 2,000 men classed as 'effectives' – i.e. not sick or wounded – were down to 806. Mountbatten was furious at what he described as Stilwell's callous indifference to the plight of the Chindit force, largely to satisfy his own quest for success. He demanded they be pulled back and eventually they were, although once resupplied and replenished they continued to operate throughout 1944 and into the final stages of the war.

Their operations were to run parallel to those of Mountbatten's former associates in Combined Operations whom he had by then installed around him: old friends from the units he had helped to create back in England. The first Commando units were sent to join his operations in south-east Asia almost immediately he took up his appointment. They were the advance party of the entire No. 3 Special Service Brigade consisting of Nos. 1, 5, 42 (RM) and 44 (RM) Commandos.

Early in the New Year, Mountbatten also confirmed his commitment

[1] Fully described in John Parker's *The Gurkhas: The Inside Story of the World's Most Feared Soldiers*, Headline, 1999.

to clandestine raiding parties by forming what he called his Small Operations Group – SOG, or more commonly known as Soggy by those who were in it. It consisted of three troops from the SBS, along with Detachment 385 from the Royal Marines Commando assault troop, four parties from the Combined Operations Pilotage Parties (COPPs), two of the RN Commandos for beach landings and four Sea Reconnaissance Units (SRUs). The last was a relatively new organisation, originally formed to provide long-distance swimmers for specific reconnaissance missions, and developed into Britain's first underwater swimming unit using fairly primitive oxygen breathing apparatus. Initially, they were all to be involved in raids down the Arakan coast and out on the islands and then, in the grander scheme of things, were to return to Burma and then on to Malaya and Singapore as the campaign picked up speed in the early summer of 1944. The SBS was also involved in secret missions into Siam and Indochina.

In fact, SOG made a significant contribution to the Allied effort in the Far East, in spite of a shortage of gear and leaky landing craft, and in the 13 months from formation to the end of the war the group was involved in more than 160 raids and operations, some under its own auspices and others as part of larger combined attacks during the final phase of the reoccupation of Burma.

To assist the running of these operations, Mountbatten also brought out the hero of Operation Frankton, Blondie Hasler, by then a lieutenant-colonel, and the redoubtable Royal Marine colonel, Humphrey Tollemache. In the end, most of these men would, if they had the chance, have said: 'Whatever did we do to you to deserve that?' The conditions were, of course, appalling in every respect: fighting an enemy capable of every nasty trick in the book, in steaming jungle conditions, torrential rain, wading through swamps and rivers, battling killer insects, mosquitoes and subsequent disease that felled so many troops.

The British Commandos assigned to Burma were now operating under a new title, 3 Commando Brigade, which eventually became one of the great fighting names of the second half of the twentieth century. They were eventually to join the troop formations being moved into the Arakan and on towards Burma for the new offensive in the later months of 1944, and the Commandos were angered by the news, relayed to them on arrival, that because of the shortage of landing craft and air support they would be used initially only on relatively small operations. Mountbatten could only promise them the opportunity of jungle penetrations on similar lines to the Chindits.

Unfortunately for them, the brigade arrived at a point when the Supreme Commander was just about to change the rules of engagement.

Until Mountbatten's arrival, war seemed to run with the seasons, i.e. that during the monsoon period, from May to the end of September, there were only random clashes. The belief that it was virtually impossible to stage full-scale battles under such conditions was shared by both the British and the Japanese. Thus, the war more or less shut down during those months. Even before he left Britain, Mountbatten had technical experts working out ways and means in which he could keep the armies of the east on the move during the monsoon, and soon after his arrival he informed his generals to fight on, regardless of the weather. He produced the evidence of the strategists that the dangers of fighting in the monsoon were more than made up for by the element of surprise. This, incidentally, was wholeheartedly supported by Field Marshal Slim, commander of the 14th Army, who wanted to get his men on the move. He demonstrated that it could be done in the coming months when the tide was at last turned against the Japanese, who by the end of the summer were in retreat.

Nos. 5 and 44 (RM) Commandos were soon to experience the awfulness of fighting in the jungle and against an enemy the like of which no British fighting units had ever experienced. Archives of 3 Commando Brigade contain the men's reactions after only a few short engagements. The 5th Indian Division had just captured Maungdaw, 50 miles from the Indian border, from where they were moving out to confront the enemy: 'Even in our short experience, we have had played against us all the tricks you read about in pamphlets: snipers tied to treetops, moving a wounded man into a field tempting you to come and get him, shouting in English, making a noise in one direction and coming in from another. Troops must be prepared for all this or they will be caught on the wrong foot . . . you are fighting a genuine fanatic who fires and expects no quarter.'

As the various British amphibious raiders began probing the Japanese positions, the enemy commanders soon became wise to the possibility of clandestine reconnaissance missions and the seaborne landings by the Commandos. Soon, they were inserting spiked bamboo stakes underwater, 275 metres from the shore, which were virtually invisible at night but could – and often did – seriously injure the small-party raiders. And so it went on for the next year or more. From these frightening beginnings, 3 Commando Brigade hurled themselves into attack after attack, and fought back the Japanese counterassaults as gradually the Allies pushed the Japanese further south and east to a point where some of the most bitter battles of the new offensive were fought. Every one of the Commandos was involved at some point or other, every one suffered the harshness of the terrain, the terrible conditions, the monsoon and the sickness. Most of all, the elements that were found to be most frightening when they first arrived remained so until the end.

This is amply borne out by the missives back to HQ and to their base in England from the officers engaged. One from No. 5 wrote of the crucial Battle of Kangaw in January 1945: 'I've never been so bloody scared in all my life. It wasn't so much the fighting as the shelling . . . they had 20 guns firing at us all the time.'

It was one of the most hard-fought battles as the war edged closer to conclusion in 1945, and especially around a feature known as Hill 170 which the Commando brigade had taken and had been ordered to hold. They defended it with their lives against a tremendous Japanese onslaught. In one section, just 24 men from No. 1 Commando defended their position against 300 Japanese. Led by Lieutenant G. A. Knowland, the survivors of a violent assault managed to hold on to their position for eight hours. Knowland himself, leading from the front, was seen firing mortar from the hip, killing six Japanese with his first bomb. When he ran out of bombs, he dashed back through enemy machine-gun fire to get more. As there weren't any, he snatched a Tommy-gun from a wounded soldier and stood bolt upright spraying the enemy, who were by then just nine metres away from him. Although badly wounded, Knowland killed a further ten before he fell back into a trench. For this action, and the inspiration to his men, he was awarded the Victoria Cross.

Veteran of the Commando movement and Deputy Commander of 3 Commando Brigade, Lieutenant-Colonel Peter Young wrote after they came off that hill and its surrounds:

> Jap dead were interlocked with our own in a proportion of at least 3 to 1. The back slopes of the hill were thick with victims . . . It was a real epic. I never saw dead so thick. The boche could not have stood five hours of it. Several 1 Cdo soldiers were found well forward, dead, in the middle of Japs having pushed on in these counterattacks after their first section had finished. One RM officer killed 4 Japs and got 4 wounds . . . all this in about two acres of ground . . . I'm convinced that no British troops ever fought better than ours on that day.

Only later did the enormity of what they had succeeded in doing become clear. If the Japanese had secured Hill 170, they would have cut off the supply routes to the British troops in this vital stage of the battle for the Arakan. There were congratulations all round for the brigade and awards galore, in addition to Lieutenant Knowland's VC. Mountbatten was overjoyed with their success and made sure that the Commando operation received lots of publicity. He, too, was embarrassingly generous with his praise and said that if the Commandos had only completed that one single operation during the whole time they were under his command it would have been worth it. Of course, they had been kept busy throughout and remained so to the point that Burma was won back and it was Malaya next stop – and then the Japs surrendered. But, as Mountbatten predicted as he left to witness the liberation of Singapore, the story was far from over . . . and in the supposed peace the Commandos would be back in the region ere long.

Elsewhere, the Commandos were running right up to the wire in other battle fronts that were the last-chance zones of German defiance. The push from the Normandy beaches was by the autumn of 1944 moving ever outward. French tanks led the advance into Paris on 25 August, but the road to Germany itself was still paved with problems. In what was a dramatic attempt to 'end the war by Christmas', the 1st British Airborne Division, with American backing, was given the task of capturing the bridges at Arnhem to give the Allies control of the lower Rhine. It was a grave miscalculation on the part of the war managers, and vital intelligence that a crack German Panzer division was resting nearby was simply ignored, discounted by Montgomery himself. Consequently, the paras were slaughtered at that famous battle of the 'bridge too far'. After eight days of terrible fighting, only 2,400 of the 10,000 men who were airlifted or parachuted into Arnhem came back.

Access to supply ports in north-western France and the Low Countries thus became crucial. As the great columns of the Allied armies lumbered along, their trucks, tanks, Jeeps and guns needed to be serviced, replenished and fuelled and the men fed. Until there was access, the supplies had to come in by road from the Channel ports, and with the front-line troops speeding ahead at an unexpectedly fast pace, fuel supplies were a constant problem. The pressure could be eased by freedom of movement into the port of Antwerp. The city had been ringed by the 2nd British Army by the second week of September, but to reach the port itself shipping had to negotiate an 18-mile cruise up the River Schelde, the mouth of which remained heavily fortified by the Germans. This was especially true of the protruding island of Walcheren, which totally dominated the mouth of the river. The Germans had already taken the precaution of sending a massive influx of defending troops and artillery and flooded and staked large areas to deter para and airborne landings.

Thus, Antwerp remained virtually inaccessible by sea and was certainly too dangerous for heavy shipping. It soon became apparent to all that the only way to open up the port was through forced amphibious landings around the Walcheren coastline supported by a heavy bombardment of German defences. Churchill, studying the proposals drawn up by the Chiefs of Staff, objected on humanitarian grounds, because of the risk to civilian populations around the town of Flushing, a prime target. Eisenhower continued to press for the attack, and Churchill agreed on the basis that bombing the town itself would be avoided, and that every effort should be made to restrict the air attacks to German defences.

This in turn placed greater reliance on a speedy amphibious landing of troops in the style of a Commando raid, going in ahead of the landings of infantry and tanks. No. 4 Special Service Brigade, still on the Continent after D-Day, was chosen for the task. No. 46 (RM) Commando, by then down to fewer than 200 men, was sent home and replaced for the operation by No. 4 Commando. With additional support from contingents of Belgian and Norwegian Commandos, three of the Commandos would land on the western side of the island and No. 4, with two troops of French soldiers, would make their landing at Flushing. They would be followed in by 155 Infantry Brigade and elements of the 7th Armoured Division, and all would receive naval artillery support as well as landing craft now equipped with multiple rocket launchers and antitank guns. Additionally, the RAF would embark on a limited bombing of the island and Spitfires would strafe enemy positions. There were plenty of targets. By the time the operation was ready to be mounted on the night of 31 October 1944, the Germans had assembled a formidable force of manpower and guns as well as their now-familiar pillboxes and strongpoints constructed from thousands of tonnes of reinforced concrete. They were virtually unassailable except for running in close and hurling grenades into the narrow slits for the gun barrels.

Potential beach landing areas were heavily mined, booby-trapped and staked, and additional hazards were sand-dunes that had been ploughed to make them difficult to manoeuvre and the flooding of vast areas of the flat landscape. The Germans were also aided by the weather: dense cloud over the whole area which restricted the amount of air support the RAF could provide. Advance recce parties went in first to establish the landing sites and guide the main body of Commandos forwards. This was achieved with few casualties, and in fact a number of prisoners were taken. But the arrival of strafing Spitfires soon alerted the Germans to trouble, and very soon their big guns were blazing even before the landings at both sites were complete. Some of the parties coming ashore took bad hits and landing craft were sunk.

The arrival of the first landings of the 155 Brigade provided the No. 4 with needed backup, and by mid-afternoon most of the first-day objectives were met. At the other end of the island, the three remaining Commandos, landing slightly later, had an even worse run-in, with six landing craft hit. Once they were ashore, the German guns opened up with terrifying might, yet the Commandos managed to create the beachheads that would allow the flow of troops ashore.

The coming battle was again ferocious. The Germans' carefully pre-registered firepower was deadly accurate, although it was soon matched by the Allies' own naval and air bombardment. On the ground, a maelstrom of indescribable noise, flames and explosions erupted as the artillery and gunners pounded away. The troops themselves faced six solid days of hard fighting that eventually came down to house-to-house battles ebbing and flowing in the midst of a civilian population who basically had nowhere to hide. One night, two British soldiers cut off from the rest of their troop were given refuge by a local family. The Germans, coming up later, discovered this and herded the whole family of five outside and shot them at point-blank range.

In spite of their defences, the Germans were on the run by the fifth day. On the eighth day, they were waving the white flag and Walcheren and the vital approaches to Antwerp were secure. It was estimated that this one action affected the fortunes of 40,000 German troops strung out in the remaining battlefields of the Low Countries and on into Germany itself. By the end of the month, after the minefields had been cleared, the port was alive with Allied activity, bringing in men, equipment and supplies for the onward march.

It was undoubtedly a remarkable achievement, far less costly in terms of the tragic and unnecessary loss of life at Arnhem, yet a very considerable boost for the Allied advance. It is exactly what could have happened at Arnhem if the top brass had acted on their intelligence reports and put off the assault until the resting Panzer division had moved on. For No. 4 Special Service Brigade, the Walcheren operation was a major success, but the cost was heavy. Already depleted after their assignments for the D-Day landings, Walcheren claimed a further hefty slice of their manpower: 103 killed, 325 wounded and 68 missing during just eight days of fighting. The brigade was temporarily pulled out to re-equip and take in reinforcements, but half their force went back to Walcheren in case of an attempted German return. The remainder, plus No. 47 (RM) Commando and three troops from No. 48 (RM) Commando, were sent straight away to join the British 1 Corps as backup against the short-lived Ardennes break-out attempt by the Germans in December and then came under the command of the 1st Canadian Army tasked with clearing the Germans out of northern Holland. In the remaining months to the end of hostilities, the brigade carried out no fewer than 28 fighting patrols and 14 reconnaissance missions.

In December all four Special Service Brigades – as already noted with 3 Commando in the Far East – were renamed Commando Brigades. The former No. 1 SSB became 1 Commando Brigade under Lord Lovat's successor, Derek Mills-Roberts, and was destined to set off for Burma after being reinforced following their return to the shores of England in September 'stinking and filthy', as Henry Cosgrove described them. Even as they were gathering their gear, 1 Commando, soon to be rejoined by No. 46 (RM) back from loan duties with 4 Commando, was switched across to western Europe – to begin what became one more remarkable episode to add to the wartime exploits of the Commandos: a long haul, mostly on foot and fighting all the way, from the heart of Holland to the Baltic.

En route, they took on a variety of tasks ranging from strengthening defences at Antwerp against possible countermeasures, then onwards to join the line at Maas, marching across Holland and spearheading the break-in to Germany, taking in four major opposed river crossings, including the Rhine and the Elbe, and ending up taking the surrender of Field Marshal Erhard Milch at Neustadt, whose arrogance earned him a swipe over the head from Mills-Roberts's baton. Henry Cosgrove went all the way with No. 45 (RM) Commando:

> We'd left Tilbury docks in a derelict old ferryboat that was once used for the Isle of Man years earlier. It brought us to Ostend, then we went by train to Helmond, in Holland, with the 15th Scottish Assault Division, moving now on foot to Massbracht, and on to Brachterbeck, which was captured with few casualties. We hit trouble a bit further on, on the banks of the Maas, and a fair old fight developed. The Jerries had got us pinned down out in the open and we were getting heavily shot up. One of the Army Medical Corps lads by the name of Harden was a real hero, wonderful fellow liked by everybody. He was dodging about getting the wounded in, then he was shot himself. They gave him a VC.[2] As the Jerries retreated, they had marvellous ways of setting up points to hold up the advancing army and we fell for it. We got pinned down and we

[2] Lance-Corporal H.E. Harden of the Royal Army Medical Corps ran forward to attend a wounded officer and two marines, dragging them back to safety. In spite of orders not to risk his life any further, he twice went back with stretcher-bearers to help evacuate the wounded, and on the third occasion was shot through the head. He was awarded a posthumous VC for his bravery.

lost quite a few men there. Eventually, the lads behind us laid down a couple of smoke screens for us to get out of it, but as soon as the smoke screens came down the Germans opened up with everything they had. They knew the range, so we couldn't move. Eventually, darkness came and we were shifted out of it.

By then, we were up in the front of the 2nd Army with the intention of carrying on through Holland and up to cross the Rhine [at Wesel]. It was freezing cold, snow on the ground, and we used to burrow into it at nights to keep warm. En route, I acquired a pet duck which I called Hector. I judged it would be a meal sooner or later. It was a bloody nuisance; we'd no sooner got near the water than Hector started quacking, and I had to tie his beak up. I had him strapped cross the top of my pack. We set off, Hector with his beak tied, and we got across. There was a bit of a schemozzle when we got to the other side but we managed all right. We got ashore and with a fair bit of firepower we edged our way in. Meanwhile, Hector laid an egg so I had to rename it Hectorine. I gave it away to some Romanian children. They were starving, poor kids, bloody starving. The next day all hell broke loose.

Eighty Lancaster bombers dropped around 450 tonnes of bombs in a first-wave attack on Wesel on 23 March, followed by a second wave of 200 bombers dropping over 1,000 tonnes of heavy explosives. This was followed by a landing of airborne troops from the US 18th Airborne Corps and the British 6th Airborne Division. And, said Henry Cosgrove, those on the ground could spot the difference but in any event were overjoyed to see them:

We were in a bad way by then and the message had gone out to get our wounded together, and we were going to get them back across the river. Then, fortunately, Dakotas were coming over dropping parachutists . . . Americans first. They were dropping from a tidy height and they were ending up in the river and everywhere else bar where they should be. They were followed in by the British, who came in low – our old friends from the 6th Airborne. Anyway, they reinforced us. It was only later that we discovered that one brigade of Commandos had been facing a whole division of Germans. It was no wonder we couldn't hold them. They'd got every support, mobile artillery, the lot. Here we were with light weapons. Once the bridges had been established, and the army came over, we had artillery and tanks, supported then by the well-known Desert Rats, 7th Armoured Division. We set off again and took a ride on the tanks. We sat round the turret. That was fine for a while until we started getting up near the Black Forest, then you had to move off pretty sharpish.

The next battle for 1 Commando Brigade was Osnabrück, home to a large German garrison. But their resistance was fairly mild compared with what had been expected, and the town was captured in short order, allowing the commandos to press on ahead of the main force coming up behind, hardly stopping for breath. The next major hurdle was at Lauenburg for the crossing of the Elbe. The brigade fought off the opposition dug in around a steep bank on the opposite bank and set up a bridgehead for the 6th Airborne Division and 11th Armoured Division to pass through. They then headed towards the Baltic, ending up at Neustadt. Henry Cosgrove takes up the story again:

There was a big concentration camp there. The smell was unbelievable. We just saw masses of people behind the fences, masses of them, all clinging on to the wire. All you could see was their hands, their faces and their eyes. They were all skeletal. And the smell; you can't believe the smell. They were mostly political prisoners: Belgians, French, every nationality you could think of. It was here that the brigadier, Mills-Roberts, hit Field Marshal Milch on the head with his baton over some remark or other. You could understand his feelings at the time. I was surprised when I heard he'd done it, actually. Although he was a wild man, the brigadier, oh, he was a wild man, he was also a damned good soldier. But none of us had any time for the Germans when we saw that place. There was a big pit with bodies in and there were some still alive crawling around on top in the pit. There were dead everywhere, wherever you went. A lot of the survivors had got out and were roaming the countryside. You can imagine how they felt towards the Germans; they weren't going to help matters at all. I'm afraid our unit wasn't too fussy about Germans after that lot, either. You can't believe that human beings could be like this, that they could do it to people.

By contrast, we were moved on again to Eutin, a most beautiful place, not very far from the Baltic Sea. It also possessed a huge barracks and suddenly we found ourselves with thousands of German prisoners because peace had been declared. Anyway, we were supposed to be looking after these prisoners – God knows how many there were – when there was some kind of alert: 'The Germans haven't given up in Norway. We're going to Norway.' So we started getting down towards Hamburg. We were going to get aboard tank landing ships and sail for Norway. But, thank God, we never went to Norway. We ended up in Tilbury.

We'd come home.

There was one other unresolved theatre of activity in which the Commandos were engaged to the very last – one that had employed the talents of 2 Commando Brigade for 18 months. Now under the command of Brigadier Ronnie Tod, veteran of No. 9 Commando, the force consisted of four units, Nos. 2 and 9 (Army) Commandos and Nos. 40 and 43 (RM) Commandos. Nos. 40 and 43 had continued where we last left them, engaged in their raids with the Partisans on the coastlines of Yugoslavia and Albania but principally on the Adriatic islands where, on Vis, they had provided a safe haven for Tito. There remained heavy German resistance throughout and severe fighting ensued in the manner for which the Commandos were trained and now equipped. Even so, they forced the enemy out of a number of key positions, including the island of Solta, where a large number of prisoners were taken before they could escape. Towards the late autumn of 1944, however, the Germans were on the retreat from the Balkans, and the Commandos joined the Partisans in ensuring an aggravated departure. Their tasks now included other aspects, such as repairing roads to allow the movement of transport and machinery. Furthermore, with the Red Army advancing from the north-east, the British began to get the distinct feeling that their Partisan colleagues, who were of the Communist persuasion, couldn't wait. With victory almost in view, friction between the two groups began to arise. In spite of all that had gone before in the way of joint efforts against German positions, the Partisans made no attempt to hide their political preferences, and life for the British troops became difficult at every turn. The Partisans' military commanders began to operate without reference to the British and even stopped civilians talking to the Commandos.

In the third week of January 1945, No. 43 Commando was pulled out of Yugoslavia for good and after a brief leave to re-equip and rest, they joined up with the remainder of 2 Commando Brigade and the 8th Army as they proceeded towards the final weeks of what had been a hard and difficult slog to the north of Italy in which large numbers of casualties were incurred. No. 40, meanwhile, remained active in the Adriatic, from their base on Vis, harassing remaining pockets of German resistance until early February, when they were withdrawn and brought back to the Italian mainland to join the final push north. Brigadier Tod's 2 Commando Brigade was now together again for what would go down in the Commando history as their last great battle of the war – at Lake Comacchio.

All available units were pulled into the operations that lay ahead. The brigade would get the support of various 8th Army artillery units as well as COPP2 under the command of Lieutenant Richard Fyson and M Squadron of the SBS led by the Dane, Major Anders Lassen, whom we met earlier during the exploits of the SSRF parties on the coast of France. Since then, the Dane had come through unscathed from many daring operations in the eastern Mediterranean with Lord Jellicoe's SBS, which he had taken under command after its detachment from the SAS.

Although called a lake, Comacchio was in fact a massive combination of a natural lake and a huge flooded area which was for the most part a shallow, smelly, swamp with a base of deep, slimy mud. Large areas had little more than 45 centimetres of water, not deep enough for landing craft. Recent dry conditions had added to the difficulties; in some parts the water level had dropped to as little as 15 centimetres. The whole sat plumb in the middle of the final stage of the Allies' advance north into the Po river valley and was surrounded by fortified German positions. This was, then, to be the object of 2 Commando Brigade's final major assault of the war, under the codename Operation Roast: to open the path for the troops coming up behind.

Two full units, Nos. 2 and 9 Commando, were assigned the task of crossing Lake Comacchio to secure objectives on the north and east of the lake which would take them well behind enemy lines. No. 40 Commando would hold their own line on the River Reno, flowing south out of the lake, while No. 43 would secure a long strip of terrain between the lake and the Adriatic coast. The attack began at first light on 1 April 1945 but the beginnings did not augur well. The mixture of craft assembled for the crossing consisted largely of flat-bottomed storm boats each towing a string of Goatleys, all heavily laden, and progress was slow. Heavy fire was anticipated from the opposing shores, and it came in spite of a 5,500-metre smoke screen to cover the landings. Although initially surprised by the extent of the incoming force, the Germans quickly mustered their response to the series of attacks now following on the targets set for the Commando units. In one attack, launched in the afternoon, the smoke cleared too quickly, and the troops of No. 9 Commando had to make their landing under heavy fire. They advanced with a piper playing 'The Road to the Isles', and by the time darkness fell they had secured their objective and captured 128 prisoners.

No. 43 Commando hit some murderous fire as they came towards their target area, and the leading troop was literally pinned down by heavy incoming machine-gun fire. Realising the desperate situation, Corporal

Tom Hunter dashed out to an exposed position and began firing his Bren gun at the five German gun positions lined up ahead of them. As he did so, a number of men who would surely have been cut down made their escape. Sadly, Tom took a number of bullets to his body and fell fatally wounded. He was posthumously awarded the Victoria Cross. The remainder of his unit, meanwhile, held their position until they were relieved at first light the following day by the 24 Guards Brigade.

Not long afterwards, a second Victoria Cross would be awarded. In conjunction with the main objectives north and west of the lake, four small islands had to be taken. This task was handed to Anders Lassen and M Squadron of the SBS in conjunction with an Italian unit, the 28th Garibaldi Brigade. He was to lead a number of raiding operations over the coming nights. In the early hours of 9 April, he led a patrol to a causeway at the very northern tip of the lake, leading to Comacchio town. As they headed towards the town, they came under sporadic fire which they silenced with a handful of grenades. A little further on, they came to a building from which six Germans emerged with their hands up, but as the men went forward to secure them they came under heavy machine-gun fire. Lassen himself was wounded instantly, and he ordered his troop to withdraw with himself giving covering fire. He kept it up until all his men were behind him and into their boats, and as he turned to join them Lassen was felled in a hail of fire. His gallantry was marked by the award of a posthumous VC to add to the three Military Crosses he had won during his exploits off the coast of France and with Jellicoe in the Mediterranean.

Ahead lay almost two more weeks of hard fighting for the brigade before the German forces in northern Italy and Austria finally surrendered on 2 May 1945. Lake Comacchio was among the most coveted of all the incredible total of 38 battle honours awarded to the Commandos and approved by the Queen in September 1957. They ranged across the whole panoply of war that had engaged their forces since those bleak days of 1940, punctuated by major actions recorded so far through Norway, St-Nazaire, Dieppe, Normandy, Walcheren and north-west Europe, North Africa, the Middle East, Italy, the Adriatic, Greece and Burma. They were actions in which no fewer than 1,706 army and Royal Marine Commandos lost their lives during the Second World War.

Chapter Thirteen

Here We Go Again!

In the immediate aftermath of the war, Commandos were scattered around the theatres where they were last engaged, helping to sort out the mess. No. 3 Commando Brigade remained in the Far East until the autumn to assist the Royal Navy and to be on hand for the formal surrender of the Japanese. Nos. 1 and 4 Commando Brigades remained in Germany until the summer months, joining the vast organisation of Allied military personnel engaged in the overwhelming multiplicity of problems left in the wake of the defeat and devastation of Germany. They ranged from rounding up Nazi war criminals and officials who would stand trial, questioning, marshalling and feeding the thousands of prisoners and attempting to assist in the restoration of some form of social order. No. 2 Commando Brigade, which had remained in the fighting to the very last, returned to England battle-weary in June 1945.

Even as they settled down to some well-earned leave, the whole future of the Commandos was being discussed at a high level, along with the arrangements for the demobilisation of the thousands of British soldiers over the coming months. For the army Commandos, the result of those deliberations would mean disappointment and anger. In September 1945 the Chief of Staffs Committee agreed to the resurrection of the 1924 recommendation of the Madden Committee that in future all amphibious strike forces should be the responsibility of the Royal Marines. The army Commandos, whose units had forged the creation of Britain's first wartime amphibious storm troops, were to be disbanded immediately and the volunteers who manned them would be either demobilised or returned to their original regiments.

The order included the SAS, although it would be revived later for the colonial wars erupting across the former British Empire in the 1960s. Other Commando-linked units were also to be dissolved. These included groups such as COPPs, Hasler's RM Boom Patrol Detachment and the Sea Reconnaissance Unit, although the skills and techniques they had invented and learned during the war would be retained by a small RM detachment which was based on the wartime Special Boat Section but now was to operate as part of the Royal Marines – eventually to become the renowned SBS.

The air of great despondency and gloom that fell over the army Commando organisation was summed up by Bob Laycock when he broke the news to his men:

> I am today more moved in speaking to you than ever before, for my emotions are not now those which I felt when I spoke to you in the past, the inspiration of battle and the exhilaration of coming danger, but they are deeper and more poignant emotions and they are these. First, the emotions of unbounded gratitude which I feel for every one of you who has helped make the green beret of the Commandos a symbol of bravery and honour whenever it has been worn. Secondly, I am very conscious of the great privilege which I myself feel in having been associated with you. And lastly, and most poignant of all, the emotions of sadness . . . for who can say that there is a more splendid example of endeavour than that which the Commandos have set during the dark misery which the world has just been through – the high resolve to volunteer . . . the love of adventure, and the skill and loyalty, and bravery unsurpassed which sometimes ended so tragically in supreme sacrifice of many of your comrades. This is the spirit of the Commandos which Great Britain is so justly proud. I therefore admit to you . . . it has fallen to my lot to tell you the Commandos, who have fought with such distinction, are to be disbanded.

Laycock also announced that the famous green beret would be scrapped, but this was later cancelled after the personal interjection of Mountbatten, and the Royal Marine Commandos would continue to wear them, as indeed they still do.

The new Royal Marine Commando Brigade would become the only amphibious raiding group and take a key role in a vastly reduced RM establishment. The RM Commando units were also reduced from eight to three, and the overall RM manpower was cut at the end of the war from 70,000 to 13,000. Only 2,000 of those would be at sea, as a result of the post-war cutbacks in the number of capital ships. The Commando banner would now be passed to the new 3 RM Commando Brigade, which went on to become one of the great fighting groups of the second half of the twentieth century. Wartime Commando numbering was picked up to maintain the links with those units that had served in the major battles, and thus the brigade was made up of Nos. 40, 42 and 45 Commando. The Royal Marines Forces Volunteer Reserve was also formed to provide a backup for special duties.

The reorganisation was completed with surprising speed. A new commando school was established, first in Wales and then at Bickleigh, near Plymouth. Basic training for all marines was at the Royal Marines Depot at Deal, Kent, an infantry training centre was set up at Lympstone and the new raiding squadrons were to be based at Eastney and later in Poole. New and tough training courses were designed for the commando units, based largely on the techniques first introduced by the army commando school at Achnacarry but updated after studies of the experiences of the war years. A sense of urgency emerged during the reorganisation, and there was good reason. Across the former British Empire there was disarray and dissent. The collapse of organised governments under the Axis boot left a void from which revolution would emerge in so many different forms, especially after the new Labour government of Clement Attlee made it clear that it wanted to disengage cash-strapped Britain from its military commitments around the world. The impending chaos was already demonstrated in Singapore, which was devoid of any form of social structure. Before they returned to Britain, 3 Commando were used, along with the parachute brigades, to set up police forces; some officers even found themselves inaugurated as temporary police officers and regional administrators. 'It was a Gilbert and Sullivan sort of place,' said Jim Absalom, who found himself installed as a police superintendent. 'We did our best to establish law and order, but there were all kinds of local organisations and groups, and if a shot was fired at night they all shot at each other.'

Worse, in the eyes of the West, was the rise of Communism. The new expanded Soviet Empire dominated northern and eastern Europe after the Allied carve-up. Almost immediately, Joe Stalin's mob of political gangsters began their menacing rise, along with the purges and mass murders throughout Soviet Russia and its newly acquired territory that dwarfed even Nazi genocide and the Holocaust. In the Far East, Mao Tse-tung's Communist army, which began its 'long march' to power in 1935, was edging closer to victory over Chiang Kai-shek's nationalists now that the Japanese had been atom-bombed out of the war. The much-vaunted domino effect, denied by some British politicians, would soon be evident with Communist uprisings across the Far East, notably in Malaya, Korea and Vietnam.

In the Middle East, the dust had barely settled before the Jews and

Arabs resumed hostilities, put on hold during the war. Palestine, then, became the first real focus of the newly formed 3 Commando Brigade. Their partners in so many past – and future – operations, the Parachute Regiment was already *in situ*, having been sent at the end of 1945 as the trouble between the two sides flared into violent confrontation. The tensions surrounding what was the Holy Land to Christian, Jew and Muslim, over which Britain was granted a mandate following the carve-up of the Ottoman Empire by the League of Nations after the First World War, had not been eased one bit by the war. Indeed, they were now exacerbated by the thousands of European refugees seeking a home in the promised land which the Palestinian Arabs claimed had been promised to them. After the First World War Britain introduced a two-faced solution: the government pledged Arab independence from the Ottoman Empire but it also confirmed that the British favoured the establishment of a Jewish 'national home' in Palestine without violating 'the civil and religious rights of the existing non-Jewish communities'. So the scene was set for the battles between the two sides which were to last for the rest of the century.

After the Second World War, the Zionists naturally wanted the British to allow the immigration of the vast numbers of Jews in Europe who were displaced, homeless and emotionally fired up by the discovery that six million of their brethren had died in the Holocaust. The 'national home' promised after the previous world war by Lord Balfour now became not merely a place to rest their heads for these lost souls, but one that was also a profound and imperative ingredient – the focal point – in the spiritual and psychological recovery.

By the end of 1945 refugees, encouraged by the Zionists and other powerful Jewish groups, were pouring into Palestine. The British government tried to limit the numbers, in spite of calls from the Americans to relent. With thousands of Jews sailing towards Palestine in ships from Europe and elsewhere, Britain's answer was to send troops and gunboats to keep them out. Thousands of soldiers were moved in and immediately faced the wrath of Jewish armed units, extremists such as the Stern gang and Haganah, the self-styled Jewish defence group. Open warfare developed between the Arabs and Jews, with the British caught in the middle. In July 1946 the British military headquarters at the King David Hotel, Jerusalem, was blown up by delayed-fuse explosives planted by Zionist terrorists, killing 94 and wounding 52. In November eight British soldiers were killed when their Jeep was blown up in Jerusalem. In January 1947 British families were evacuated from the region. In April British police barracks near Tel Aviv were blown up.

In the meantime, one of the most dramatic Jewish attempts to break the British immigration embargo was already on its way. An old US troopship previously named the *President Garfield* and now called the *Exodus* left Baltimore manned by an American volunteer crew. It picked up its cargo of 5,000 Jewish refugees waiting at a small French port near Marseilles and reached Haifa on 18 July 1947. There, as the packed ship docked, she was confronted by a boarding party of British troops which the refugees attempted to deter by hurling tins of food and fighting with iron bars and tear gas. The hand-to-hand battle lasted an hour or more, before the troops gained control and herded the refugees on to three other waiting ships bound for Cyprus.

By then, the British had detained more than 20,000 would-be Jewish immigrants and transshipped them out of Palestine to holding camps in North Africa and Cyprus. In the terrorist reprisals that followed, two British soldiers, Sergeant Mervyn Pace, 43, and Sergeant Clifford Martin, 20, were kidnapped by Zionists and were discovered hanging from two eucalyptus trees in a suburb of Haifa, with notes pinned to their shirts stating they had been executed as spies. Many other atrocities were committed before Britain decided to place the whole dispute before the United Nations. There, the politicians bickered and stalled. Britain said it was pulling out anyway, with or without UN intervention.

The UN Special Committee did not reach Palestine for its inspection tour until the early summer of 1947. They were boycotted by the Arabs, who could not even contemplate a glimmer of cooperation. Predictions of a bloody catastrophe abounded, and at that point the Commandos were sent in. George Richards was still a teenager when he joined 42 Commando in 1946. He was among the intake of new recruits who went through the newly established training school, and he, and many like him, were soon to see action. In 1947 3 Commando Brigade was moved out to Malta, based there as a ready-to-move outfit if and when the need arose.

> We were shipped across on a couple of occasions to Tripolitania, which at that time was in the throes of being handed back to the Arabs, but we had to quell a few riots that occurred between rival factions, then the whole brigade was put on standby for an immediate move to Palestine. The commando was given 12 hours' notice to move, which meant rushing around and packing equipment. We embarked for Palestine on the LST [Landing Ship Tank] *Striker* and the *Dieppe* to sail across. We hit some very rough seas on the way over; the great doors banged and rattled and the flat bottoms of these vessels made them rock and bounce.
>
> Eventually, we ended in Haifa at the alert. All guns were manned.

It was a landing under arms because at this time the Jews had really got their act together. We landed at Haifa and my Commando was billeted in Haifa itself. We took over an empty hotel and set up our guns on the roof. I had a wireless station up there. Our tasks were generally to put ourselves between the Jews and the Arabs, running security patrols searching for terrorists and weapons. It was while we were in Haifa that the Jewish vessels were trying to land their illegal immigrants, and threats were made to kill our CO, Lieutenant-Colonel Riches, after we stopped two or three vessels from landing. So, he had to have a bodyguard everywhere he went. We stayed at Haifa for about a month and then suddenly the whole brigade was ordered to move to Jerusalem.

We set off by road in convoy, and a curious confrontation occurred. Just outside Rām Allāh we were stopped by the TransJordan Army, which was commanded by English officers. They wouldn't allow us to go any further. They had great big field guns and everything mounted. They were controlling the road to Jerusalem. It took quite some time for our commanding officer to parley with them, but eventually they allowed our convoy to pass through their lines. We finally reached Jerusalem and set up in different areas; 42 Commando was at a village settlement just outside the city, in an old nail factory.

We used this as our headquarters. In the distance there were three Jewish settlements and on the opposite side of us were two or three Arab settlements. So there was the potential for trouble, which occurred when, for example, a woman in one of the Jewish settlements would come out and start hanging out her clothes; then the Arabs would open fire. Any movements like that created fire from the Arab settlements and vice versa. It happened nearly every day. They'd start firing at each other and sometimes hit our place as well, accidentally. It was all small arms and machine-gun fire. It was wild, very erratic, and our job was to try and maintain peace, obtain a truce for them to stop firing. But no sooner had it stopped than it would start up again. This was how it was going on.

The increased fighting was evident from the statistics. In one six-week period monitored by the UN in January 1948, nearly 2,000 people were killed in Palestine: 1,069 Arabs, 769 Jews, 123 British and 23 others. In the end the UN recommended the formation of an independent Jewish state by the partition of Palestine. Britain was, in the end, pleased to accept the UN verdict and leave as quickly as possible; it had much bigger problems confronting it with the withdrawal from India, where a similar solution was being pushed through by Mountbatten to form the independent states of India and Pakistan and the beginning of some of the worst ethnic violence of modern times in which four million of the population were to die. The Palestinian 'solution' led to a dramatic upsurge in the violence as each side jostled for position and territory. In March British troops used their biggest 25-pounder guns to shell Arabs dug into hillside position 20 miles west of Jerusalem, resulting in 100 casualties. In May the British began pulling out its remaining troops ahead of the birth of Israel, set for 14 May. George Richards remembered the scenes in the period leading up to the deadline:

> We had been moved into Jerusalem itself. The Palestine police force, which was largely British, had been pulled out of the city, and we were to go in to replace them and, it turned out, to destroy their barracks. It was quite a modern barracks, but they'd withdrawn the police force by then. Anyway, we got set up in the barracks, and on corners there were armoured vehicles dug in with heavy machine-guns on the perimeter, which was surrounded with barbed wire. We found that in Jerusalem at night, it was just a case of everybody shooting at anybody. There were night-time curfews, but as soon as dusk fell the shooting started. They were all on flat roofs, if you can imagine, firing at each other. Really, I don't think anybody knew who they were firing at in the darkness. There was really just one heck of a row all the time. You'd hear bullets ricocheting around our area. One particular night they made a terrific raid against us. We found out later they were Haganah people, presumably looking for arms. The machine-guns opened fire and there was quite a big fight that night. Next morning there were quite a lot of Jewish Haganah people with their blue and white sashes killed on the perimeter. A lot of them were girls. It was a tragic sight, just young girls, all fighting for Haganah. That was all night long. We were fighting back, defending what we had there.

Two days later the Commandos were ordered to destroy everything in the barracks that could not be carried. The evacuation that everyone had expected might come was underway. Richards again:

> All the equipment that was inside, much of it brand-new, big diesel engines for power and things like that, we had to smash. A lot of stuff was simply burned. Then the whole unit was formed up into a huge convoy, miles long. It had all the 1st Guards Brigade, the guards' tanks, the Royal Engineers and all other forces pulled in for the withdrawal. The whole convoy was lined up and set off from

> there to go back to Haifa. Aircraft flew overhead; Spitfires darting backwards and forwards guarding it from attacks. Any truck or vehicle that broke down was just pushed over the side and set on fire. Behind our truck, which was a three-tonner, was a guards' tank, and as it came down the hill its brakes wouldn't hold. It kept sliding towards us and we kept seeing it coming ever closer. Eventually, it hit our truck, which meant we had to get out, push it over the side, set fire to it and get into another one. This is how it went all the way along. It was such a huge convoy that there were loads of vehicles that had been discarded.

Elsewhere, the opening shots were fired in other troubles that would engage the British Commandos and a considerable portion of the British military might for the next five years. The Far East was already on a knife edge as Mao Tse-tung's army continued its rout of the Chinese nationalists in spite of massive American aid, and in a knock-on effect Communist leaders who had been allies of Britain during the war had turned their forces, trained with British help and arms to fight with the Japanese, against British interests in Malaya. On 22 June, less than a month after the pull-out from Jerusalem, the British rounded up 800 known Communists following the murder of a number of rubber planters. Britain would soon be pouring troops into the troubled country as Communist terrorists, mostly from the Chinese population, began widespread attacks. The possibilities for trouble in this volatile region were endless, and the Commandos would soon be pulled into it, along with numerous other elements of the British army. But first, 3 Commando was required in Hong Kong for a spot of gunboat diplomacy as Mao's army marched on. Officials in the British colony of Hong Kong were increasingly alarmed at the arrival of the Communists, virtually on their doorstep, and, additionally, thousands of nationalists fleeing from Mao were expected to be clamouring to reach British colonial soil.

Among the newest recruits in the party was Jim Green from Ormskirk, Lancashire, who not long before was working on a farm driving tractors until, at the age of seventeen and a half, he took himself off to the Royal Navy recruiting office to join up but discovered they wanted only stokers. He told them he wanted something on the open decks. The recruiting officer took him across the corridor and by the time he came out he'd signed on for 12 years' minimum service in the Royal Marines. On 18 May 1948 he reported to the Deal depot:

> It was a bit of a shock at first, as a country boy, but once you settled your mind to accepting that the guy with three stripes was in charge,

> it was no hassle. I quite enjoyed it, in fact; some took to it, some didn't. We went to Portsmouth for gunnery training, then fourteen weeks' infantry training, then to Bickleigh for six weeks' Commando training. I was about to begin a specialist training course when the panic in Hong Kong arose and we were told to prepare to go straight out. There was a huge draft, five or six squads, and we were to join brigade base at Malta and went on board a liner requisitioned specially for it. We were dropped off at Malta while the ship went to the Middle East to pick up the other Commandos and then it came back to pick us up for our journey onwards to Hong Kong. I joined 42 Commando stationed at Whitfield Barracks at Kowloon and went straight into duty. We were on border patrols in the New Territories, guarding ammunition dumps and the governor's residence. The border patrols were to stop refugees getting across from China. There were quite a lot getting through. There were four rifle troops, A,B, X and Y, so named because they corresponded to the gun turrets on a cruiser. Then there was S Troop, support troops with heavy weapons – Vickers, mortars and so on – and then there was a headquarters which dealt with administration. It was a terrific posting for me. The furthest I'd ever been from home was Liverpool. It was wonderful, and as it happened there was very little trouble that we couldn't handle.

By December 1949 the civil war in China was reduced to a few sporadic outbreaks of fighting. The Communists were firmly established in Beijing, where the nationalist troops lost 600,000 men and the government shifted its capital to Taipei on the offshore island of Formosa (later Taiwan). The domino effect, however, was already in progress. North Korea had already declared itself a republic, and the French were facing increasing troubles in Indo-China, later to be known as Vietnam, having outlawed the Communist leader, Ho Chi Minh. In the same month, the British government ordered four regiments into Malaya in the biggest onslaught yet against the Communist terrorists operating in the dense jungles. The rebels were calling themselves the Malaya Races Liberation Army, but it was still run by Ching Peng, a Chinese Communist leader who was awarded an MBE by Britain for his services to the Allies in the Second World War and who, apart from collecting his medal from King George VI, had marched in the 1946 London Victory Parade. Ching had amassed a well-trained and disciplined army of 5,000 men who had consistently eluded British forces since the Malayan 'emergency' was declared in 1948. By March 1950 the Communist Terrorists (CTs) had taken a hefty toll: 863 civilians, 323 police officers and 154 soldiers.

Their own casualties ran to 1,138 killed and 645 captured. Elements of 3 Commando were already serving in Malaya, along with other British military units, and the brigade had set up headquarters at Ipoh. A very young Jim Green was with 42 Commando when they were shipped out of Hong Kong:

> We moved to Malaya in early 1950. We went to Penang island first to get jungle training, learning ambush patrols, anti-ambush drills, setting ambushes and tracking. Acclimatisation was important, given the conditions. It was all very new to most of us. I learned how to become a scout, working well ahead of the main troop. It was a dangerous job but surprisingly quite popular with the men. I certainly enjoyed it. You had to be very alert and you saw much more, like wild animals which would have dispersed by the time the main troop came up. You also had the possibility of running into the ambush first, of course, but that merely kept up the adrenaline. There were two scouts, a leading scout and a second scouting, and in between them a Borneo tracker, very loyal little chaps and proud to serve with the British army. They were excellent trackers: a disturbed leaf, a footmark on a jungle floor – they'd see it and be able to tell us what it was. The Bren group, meanwhile, were bringing up the rear: so there'd be ten in a section on patrol – three in the Bren group, seven in the rifle group, of whom two would be scouts.

Their first billet was two wards of the Central Chinese Mental Hospital. It consisted of a series of brick walls built to waist height with mesh grilles up to roof height. They didn't stay there long, however, because the sewerage systems for the inmates were open drains and all the waste was just swept into deeper levels until it reached a pit close to the Commandos' quarters. The stench was appalling. There were only about a dozen out of Green's whole troop who didn't go down with dysentery. So they moved out to a tented camp in a rubber plantation and were based there for some time:

> Each troop was allocated quite a huge area to operate in. The troops would have a troop commander, and two half-troop officers would take half a troop, i.e. two sections. On normal patrols there would be two section patrols, or, depending on what the range was, the sergeant might be in charge. In my half-troop we had a chap called Ensleigh who was a nephew of 'Mad Mike' Calvert. He hated being in base camp and consequently we were always out on patrol. We

did one patrol lasting seven or eight weeks[1] passing through three states in Malaya, simply because he had this idea that the terrorists moved about so freely they must have a particular route, possibly a highway, which they used. His aim was to try to find them and throw some ambushes on them. Some of the areas we were on were unmapped – not charted at all. We'd set ambushes on bits of tracks we'd find. We'd hang grenade necklaces, but more often than not all we ever trapped in them was a wild pig. The necklaces consisted of 36 grenades with the pin of each one tied to a string attached to a tripwire; if something kicked the tripwire it pulled all 36 pins out and they all went off within seconds of each other – the idea being that the lead man of a terrorist group would hit the trip and the guys coming up behind him would be in the ambush zone. That was the hope of it.

We were supplied by airdrop with fresh food and clothes. Supplies were ordered by radio and came straight down. It was hard work, but as a young fellow of 20, 21 I thoroughly enjoyed it. But life wasn't all patrolling. We'd work with jungle police going into villages that were thought to be harbouring terrorists, and we would go in to round them up. It might even be a two-troop operation. You would take up positions all round the village and then a party would go in and rouse out all the inhabitants, males, females separated in lines. The informer would be brought in under cover and he would go down the lines and point out anyone connected with the terrorists and they would be arrested and taken to the local police station. Sometimes, to reach the villages by first light, we'd have to turn out at two or three in the morning. From where our camp was we went over a bridge over the river which was made of railway sleepers which rattled when the trucks went over. You could hear this through the silence for miles. It was only sometime later that we learned that villagers nearby knew every time we were turning out; consequently, we avoided the bridge and crossed the river itself. It was only years later, at a reunion, I saw photographs of the river and it had crocodiles, although we never saw any. Leeches, yes. When you came through the river you always had to check, although you'd see anyway because blood would be pouring out of your jungle boots. I once burned 32 off my legs with a cigarette in one go.

We also discovered later on that our galley boy and our police interpreter were both informers. They would know when we were

[1] Later to be adopted as regular practice in jungle patrols, lasting in fact up to 12 weeks in the Borneo confrontation.

going out the next day, because if the patrols were going out they would be rationed up the night before. The galley boy would know every move we were making. On our first patrols in the early days we were finding our way around. Our two sections were working two separate ridges through the jungle and we'd gone through a clearing when we were about to enter the jungle proper and the boss told us to take a break. As we were sitting having a smoke, we saw three guys coming down the other ridge, running. We thought it was the other section, but the boss put his glasses on them and realised they were terrorists. We opened fire and shot and killed two; the third was wounded because there was a blood trail, but he got away. Now, the two we killed – one was identified as a Japanese armourer, presumably a Jap who'd stayed after the war, and the other was a known terrorist from that area.

The routine for identifying them if you were some distance away: you didn't carry the whole body back; you cut the head off. None of our guys were rushing forward to do it. So the tracker from Borneo was asked if he would do it. He was overjoyed at this because he came from native head-hunters; he thought he was being given the head of the terrorist as a trophy, and having cut it off he didn't want to give it up. The boss wanted to put it into a bag that we carried and there was quite a little tussle going on.

The Japanese, when they left, had buried a lot of their ammunition and the Chinese, who had plans for terrorist activities, went around recovering the shells. This is where the Japanese armourer came in; he would be able to dismantle the shells and make home-made bombs or what have you.

The long-distance tracking patrols set up by Lieutenant Ensleigh were reminiscent of wartime operations conducted by his uncle, ex-Commando Mike Calvert, who by coincidence was in Hong Kong at that time as a British staff officer. In May 1950 General Sir John Harding, Commander-in-Chief of Far East Land Forces, asked him to make a detailed study of the problems facing troops in Malaya and to report back as soon as possible. Calvert took off for the jungle and made a 1,500-mile tour along routes infested with terrorists. He visited villages, spoke to their head people and then prepared a long summary of his findings which became the basis of the Briggs Plan. General Sir Harold Briggs was director of operations. His plan was a two-phase operation: first, to close down 410 Malayan shanty villages which the terrorists used for supplies and shelter, ruling by fear rather than acquiescence. The villagers were moved into fortified settlements where they could live and farm without fear of being harassed by the CTs. Next, the troops would embark on ambitious operations to deny food supplies to the guerrillas in their jungle hideouts. A newly created SAS-style force, to be led by Calvert himself, would perform a secondary role of making contact with the indigenous people of the jungle regions and winning their trust. Calvert set about raising his troops by posting an invitation for applications for special duties in a special force for the Malayan emergency. Many who joined were former wartime operatives in the now-disbanded squadrons of the SAS, to be called Malayan Scouts, the forerunner of the reformation of the SAS. Young Jim Green was there to witness its success:

> The hearts and minds operations to deny the terrorists food supplies certainly had a good effect. This process, as it developed through the country, gradually cut down the terrorist incidents substantially, and by the time I came back to Britain in November 1951 there were very few incidents; our patrols were finding nothing. The native population were quite happy because they didn't have to support the terrorists. In spite of everything, I thoroughly enjoyed my time in the jungle and had my brother not died that October I would have stayed on. I look back now and think I must have been mad, the things we did. But at the time it was a great adventure, slogging over the hills and mountains. The comradeship, everything was great. We've often talked about it, friends in the service, in later years, and I think it's quite true that the older you get the more sensible you are. Because in those days, as youngsters, we did some damned silly things.

While Green returned to England to resume his specialist course at Bickleigh, the Commandos remained in demand. The Malayan emergency trundled on intermittently for another five years until the country was granted independence. Not far over the horizon, the sound of fresh and bigger guns could already be heard.

Chapter Fourteen

The Horror of Korea

A special British Commando unit was raised for the Korean War which began in June 1950 and ran for three bloody years. Apart from being one of the most brutal and dehumanising conflicts in modern history, the war was unique for a number of reasons, not least that it was a forerunner of Vietnam and thus its horrors soon became somewhat overshadowed. Controversial issues beyond the battle lines included the brainwashing techniques used by the Chinese Communists on prisoners of war as well as other atrocities, and allegations that germ warfare was used by the Americans, which was never substantiated but is not beyond belief, given what happened later in Vietnam. The US did, however, make extensive use of napalm for the first time against mass populations. The Korean War also had the distinction of being the first conflict in which the United Nations had sponsored the use of military force against an aggressor, although it was largely driven by the Americans, who also supplied 90 per cent of the troops, hardware and equipment.

While the war itself and battles fought are beyond the scope of these chapters, some poignant recollections in the ensuing pages from those who were there provide a disquieting insight into the involvement of the British troops, including the first-hand accounts of brainwashing. Before going on, a brief recap of events may prove helpful.

The causes lay once again in the history of the nation. Japan had ruled Korea since 1910, but, when the Russians and the Americans kicked them out in 1945, Soviet troops occupied Korea north of the 38th parallel, while American troops controlled the country south of this line. In 1947 the UN General Assembly declared that elections should be held throughout Korea to choose one government for the entire country. The Soviet Union refused to sanction elections in the north, but on 10 May 1948 the people of South Korea elected a national assembly. The North responded by forming the Democratic People's Republic of Korea. Both claimed the entire country, and the Communists made their move and invaded South Korea on 25 June 1950. When UN demands for withdrawal were ignored, member nations were asked to provide military aid to South Korea. Sixteen UN countries, including the United Kingdom, sent troops to help the South Koreans, and forty-one countries sent military equipment, food and other supplies. The North Koreans, meanwhile, had the backing of the Soviets for equipment and advisers and the Chinese for additional manpower, of which they had a vast reservoir.

The ground forces of the UN nations, along with the South Koreans, came under the command of the US 8th Army, led by General MacArthur. US troops were the first to land, followed soon afterwards by British Commandos, and initially the war went very badly for the UN force. The North Koreans dashed south and quickly captured the southern capital of Seoul and pushed the defenders into the south-eastern corner. With a massive injection of UN reinforcements, MacArthur launched a surprise amphibious invasion in September, well behind enemy lines at the city of Inch'ŏn on South Korea's west coast, about 25 miles west of Seoul. Very quickly the North Koreans were driven back across the 38th parallel.

The Chinese entered the war officially when UN forces crossed into North Korea on 7 October and later recaptured P'yŏngyang, its capital city. By the end of October advance units of the UN force were fighting the Chinese as well as North Koreans. After initial success by MacArthur's troops, the Chinese returned to the attack, this time fielding huge numbers of soldiers and heavy armoury, and in the bitter Korean winter they soon had the UN force in retreat. The Communists advanced back into the south, and it took a massive counterattack by the entire UN command, codenamed Operation Killer, on 21 February 1951 to hold them. Under pressure of superior firepower, the Chinese slowly withdrew northwards. Heavy fighting continued, however, month on month on the ground and in the air. The UN now had more than 300,000 troops in Korea in addition to the 340,000 of the South Korean army. The Communist forces were put at close on 900,000, which included two Chinese armoured divisions and one mechanised division with 520 tanks.

Air power played a key and spectacular role in the war as the first-ever war with dogfighting supersonic jets. The Chinese had developed into a major air power. Half of their 1,400 aircraft were Soviet-built MiG-15s, operating from bases in Manchuria and seldom venturing over UN lines. The UN virtually lost its air supremacy over what was known as MiG Alley in north-west Korea until President Truman ordered the urgent supply of new F-86 Sabres. Large-scale air battles and dogfights resulted in the loss of almost 1,000 aircraft by the two sides, mostly by the Chinese.

There was great naval activity, too, with the British Pacific Fleet operating on the waters west of Korea, while the American 7th Fleet took the east coast. They also provided massive support for the marines and Special Forces as they began their forays deep behind enemy lines, attacking and destroying Chinese supply lines and attacking North Korean railway systems, bridges, electricity plants and industrial centres.

The ground force included an initial draft of around 4,000 from the United Kingdom, departing in late July and early August. The first to go were the Commandos, a unit raised entirely from volunteers, principally from among the Royal Marines and including a section from the SBS. It would be called 41 Independent Commando and was hurriedly pulled together in July under the command of Lieutenant-Colonel D.B. Drysdale, RM, and dispatched at once to the US naval base at Yokosuka, Japan.

The 41 Commando was subsequently involved in many operations, some of which are described below, along with troops from British army regiments. But such were the conditions that it is perhaps more pertinent to concentrate here on the experiences in the aftermath of battle in two stories produced below which provide a striking contrast in their eventual outcome.

Before doing so, let us remind ourselves of the appalling human cost of the conflict which raged on, ebbing and flowing in battle until an armistice was agreed in July 1953: America suffered 157,530 casualties; deaths from all causes, including illness and POWs, totalled 33,629, of which 23,300 occurred in combat. South Korea sustained 1,312,836 military casualties, including 415,004 dead. Casualties among other UN allies totalled 16,532, including 3,094 dead. Communist casualties were put at 1.9 million. An estimated two million civilians, north and south, were killed, and many million were made homeless.

Among the many British troops to find their way into this nightmare was John Peskett, from Suffolk. He had joined the Royal Marines in 1947, volunteering at the age of 17, and after basic training he joined 45 Commando. Within a year he was in Palestine in a rifle section, was then posted to Malaya and into the jungle and finally had a brief encounter with a Communist league of youth in Somalia who were threatening an RAF station in Mogadishu. In 1950, approaching his twenty-first birthday, he volunteered for the 41 Independent Commando and went to Korea in September in the first wave of reinforcements. A heavy draft was raised to replace a large number of casualties sustained in a desperate battle at the frozen Changjin Reservoir, where 40,000 UN allied troops were cut off and were forced to fight their way to an evacuation point. Literally, thousands were killed, wounded or captured:

> We were rushed out to Japan with insufficient clothes. We did training to use American weapons so that only one lot of ammunition was necessary, and we were a pretty gung-ho outfit. It has been said they combed the jails of Britain for 41 Commando, but that's a load of rubbish. We were all volunteers; nobody was pressed – some from Malaya, some from Hong Kong and some from home. It was a good outfit altogether. The American marines were an excellent crowd – a mixture of good, experienced men with wartime service and young 17-year-olds fresh out of boot camp. We made many great friends with the US marines there, and the friendship was continued after the action with reunions. We began operations almost immediately, based on an island off the Bay of Wŏnsan. From there we began patrols and raids about 100 miles behind the lines. It was a listening patrol, landing by rubber boats. We would crawl up to the outskirts of a village, listening and noting all that we heard and saw, slide back to the boats and slip away to the island. These were followed by what were called tongue-snatching patrols, where we would go ashore, ambush a North Korean patrol, snatch the officer to bring back for interrogation and dispose of the rest. The listening and tongue-snatching patrols were essential; they were the only way we could get information and intelligence because the villagers were scared to death; as they were in all of these Communist countries. They'd be slaughtered if they talked.

In June 1951 Peskett was seconded to No. 1 Section of Baker troop, staffed by men from 3 SBS who were on clandestine sabotage missions behind the North Korean lines. On one of these raids, there was an enormous storm, a typhoon, and while the others were ashore their landing craft had been kept off the beach all night. The next morning Peskett and a scratch crew were to relieve the marine they'd left with the boat and sail it to the lee of the island. He set off with Quartermaster Sergeant Day and Marines Aldridge, Hicks and Bamfield:

> We started off, and the storm was still going strong. The waves were high and the wind whipped the tops off; it was like sailing in a blanket of water. We had to go out to sea to clear the rocks, and in doing so the single engine overheated; the sand traps were full of debris. The engine stopped and it took me sometime to get it

restarted by clearing out the sand traps; meantime, we were being tossed around without power. We started off again and then one of the steering wires broke. We had to repair that by having two men hauling on the rudder. By this time I hadn't a clue where we were. We couldn't see anything because of the wind-driven sea, and we struggled to keep afloat. We limped around for several hours before I noticed we were moving into shallower water.

Then we saw a beach, with a ring of rocks with a lagoon visible beyond. There was a hole in the rocks that the landing craft would go through, but as luck would have it a huge wave took us right over the top of the rocks and we hit the lagoon. The boat was beached; we were safe and ashore. The sergeant-major detailed me and Marine Hicks to do a reconnaissance. We climbed to the top of the hill and saw in the valley below a lot of lights. It was now dark, and we decided to lay up for the night and get some sleep. At first light we went back to the top of that hill and saw that it was the encampment of a whole North Korean battalion. We had hit the jackpot.

There was only one way out – by sea. I went to inspect the landing craft and discovered that the sea had gone down considerably and to my horror the landing craft was 200 yards from the water's edge. And since it weighed 30 tonnes there was no way we would be able to move it. Marine Bamfield and I hauled the floor out and decided to see if we could build a raft. The sergeant-major agreed; he wasn't at all happy about the raft, as to whether it could do the job. It would mean two of us in the water all the time and two on top. One thing was certain, however: we couldn't stay where we were; there was nothing else we could do.

Nine o'clock came and we were ready to set off. The sergeant-major said he had to go to the toilet and disappeared into the brush. As he did so, we were challenged by a Korean patrol, shots were fired, a grenade went off and we were in the bag. The sergeant-major was rounded up and was brought out wearing just a pair of blue undershorts which caused him a bit of embarrassment. We were all then taken to a bunker in the valley and individually interrogated. Each one was told that he would be taken out and shot. They took the youngest one, Jimmy Aldridge, out first, and he was walked around the corner and a shot was fired. Hicks went next; same again. Then Bamfield, then the sergeant-major, and as he passed me he said: 'We'll tell the bastards nothing.' Another shot and that just left me. I thought, Christ they've all gone. They began to interrogate me . . . and I, too, said nothing.

In fact, they were not executions but a ploy to get each one of them to talk – which none did. They were then taken as prisoners of the North Koreans and marched to the town of Wŏnsan, which had suffered greatly from air raids. Oil containers had burst and oil was running all over the streets. The five Brits were in danger of being attacked by the local populace, but the captain of their guard held them back. They marched on to a huge camp. The date was 29 August 1951, and they were kept in a hole for several days and received meagre rations of rice. They were then back on the road, now with transport, being driven to the main prisoner-of-war interrogation centre at P'yŏngyang. They were on a truck moving slowly up the side of a mountain, all five of them chained together, when an American aircraft piloted by a then Major (later General) Gerald Fink of the US Marine Corps sent a couple of rockets in their direction, blew the truck over, killed the driver and his passenger upfront while the rest of the troops and prisoners in the back of the truck were thrown clear. If they'd been thrown to the right, they would have fallen down a 300-metre sheer drop.

None of those in the back was seriously injured. One of the guards wanted to shoot the British prisoners there and then, but the young captain in charge stopped him. One young Korean soldier had a leg injury and could not walk. Peskett told the young officer to get him some bandage; he dressed the wound and the marines took off their combat jackets, cut a couple of stays and made a stretcher and carried the wounded soldier to their next destination. They marched on to reach a village where an army doctor eventually came along. He spoke excellent English and congratulated Peskett and his comrades on their first aid:

> The villagers were so pleased that they gave us the head of a horse, which was killed in the air raid, and we ate it with a bowl of rice. We continued on foot then to another village. We were to be interrogated by an officer who spoke good English. He did not get any joy, either. By then I had developed very serious dysentery, very serious indeed. He sent the others on their journey to the interrogation centre and I was left in the village with one guard; I stayed with a family, two women and two children. Neither of us spoke each others' language. They did everything they could to help me. I stayed there for a couple of weeks. If I could go back and if I could find those two women, I would give them everything I own. They were wonderful.
>
> The older one sent my guard into the hills to find some particular herbs. She then made a mixture and told me to drink it. It was horrendous, dreadful stuff. I drank some every day and gradually it

got rid of the dysentery. I only had one pair of trousers, those I was standing up in. The younger woman used to lend me a pair of her trousers. And as I fouled one, they would take them away, wash them and bring them back. They used to take me down to the river and let me sit in the water and clean up. I couldn't shave or have a haircut, of course.

One day, the guard, a captain, indicated that there was a company of Chinese soldiers coming through the village. He put a brass bowl alongside me and sat me in the doorway of the hut, and as the Chinese came through he charged them a cigarette per man to have a look at me; none had ever seen a westerner before. We finished up with a huge bowl of cigarettes. This guard one day took me into a hut and gave me a piece of paper. It was a questionnaire. He told me to fill it in, but being a cocky little devil in those days I wrote a lot of facetious answers. He got really cross when he came back and beat me about the head with his machine pistol. I lost a tooth and had a cut eyebrow and cuts on the face and cheekbones. The guard came back with me to my hut where the two ladies were and the elderly one went absolutely spare. She picked up an iron pot they used for charcoal and she threw it at him and chased him through the village.

Eventually came the day when another man, a very smart young man, came into the village, and I was called to the captain's quarters and told I was going to another place – they didn't say where. And this new soldier would take me. I was allowed to say goodbye to the villagers. We were going to the interrogation centre to join the others. About a mile along the road, the soldier who had been jabbering away in Korean suddenly began speaking to me in perfect English. It turned out he had been caught up in the Korean army when they captured Seoul; he was at Seoul university and to keep himself alive he had joined the North Korean army. Anyway, I still wasn't very well, and he said we would go straight to an army hospital. He persuaded a doctor to have a look at me, and after the examination he said: 'You know, you are going to die.'

He asked me what I would like to eat, and the only thing I really craved was rice pudding. He said, 'That's easy' and he brought up a bowl of rice pudding with lots of sugar. I was bedded down at the end of a ward with Korean soldiers, all the wounded, and my guard, Kim, gave me a gourd and told me if I wanted to go to the toilet during the night I should take it outside into the road. Well, sure enough, I had to, and I was sitting outside on this dirt track when there was a terrible noise of aircraft and I saw two rows of purple lights coming down. Just beyond me, the bullets started to hit the deck. I was sitting between the two, although the raid was over before I realised what was happening. Kim came rushing out with my kit, and said: 'Let's go.' The planes had hit the hospital and caused a lot of damage, and I wouldn't stand an ice-cream's chance in hell once they got hold of me.

We disappeared into the night and eventually he delivered me to the North Korean interrogation centre. I was thrown into a hut. The next morning, outside, I heard an American voice: 'Is that John Peskett? Happy day, and the rest are here.' It turned out to be Gerry Fink. Having made contact, it was pointless keeping me aside and they released me into the camp. That's how I met Gerry Fink, the pilot who shot up the convoy we were all in.

The camp was not a happy place. It was called Pac's Palace after the major who ran it. He used all sorts of methods to extract information from his clients. I had been captured so long that any details I had were so out of date they would have been no use to them anyway, even if I'd told them. They wouldn't believe that. I was beaten several times, with fists and flat pieces of wood. One of my punishments for refusing to talk was to make an air-raid shelter out of a rock face. I was chipping away for days on end with a piece of metal trying to burrow into the rock face: it was just a soul-destroying exercise.

If you were judged to be uncooperative, you'd be thrown into a round pit with a lid on the top. I finished up there several times in the interrogation camp. Once, I discovered a bloke who was dead, and had been down there for some time. I spent the night trying to fend off this smelly mess. It was very nasty. It was more than likely a Korean political prisoner. They used to treat them dreadfully. We saw a chain gang go by one day, men, women and children, with chains around their necks and feet in very poor condition. They were Koreans themselves.

The interrogators and the interpreters were all university graduates, very clever people, who would try to slip in the odd question to find out things. All of us tried to dodge the questions as best we could. I was once asked specifically to draw a plan of the marines' Eastney barracks, which I did and which was totally fictitious. The problem was that none of us had been briefed about what was likely to happen if we were captured. If we had been given some pointers, we could have prepared ourselves for it. Most of us therefore were ill-treated for being what they termed uncooperative.

It was at the interrogation centre that one of our number, Jimmy Aldridge unfortunately died. We buried him one morning on the side of a hill just outside the camp. Eventually, the time came when we were to be moved north to a permanent prison camp. There we were joined by Captain Anthony Farrar-Hockley[1], who had been tortured and beaten for escaping. I saw this dishevelled figure standing in the doorway. He straightened himself up and said in a very well-spoken voice: 'How many officers are there?' I knew I was in the presence of a real soldier. By then, there was a real mixture of British army, United States air force, a couple of South Africans, a Dutchman, a Turk, a couple of Frenchmen, and we'd paired off. Our food consisted of two handfuls of meal that they fed chickens on. If it was a male soldier giving it out, we had a large handful; if it was a girl soldier, we had a small one. Sometimes we could steal some salt.

This turned out to be a forming-up place for what we were about to embark on – a long march north. The weather by now had begun to get cold. There were flurries of snow the further north we went, and having been captured in the summertime we had no warm clothing. On the way, a Korean soldier pinched my boots, heavy climbing boots. He wore them for a day and then threw them over the side of a cliff because they gave him blisters. We eventually got to a village in a snowstorm. Captain Washbrook, Royal Artillery, and Captain Farrar-Hockley were laid low on an ox cart. We were all struggling. Several had already died on the way. Walking was far better than lying on a freezing cart, but they couldn't walk and had to be carried. I think if they'd been other ranks, they'd have been shot and left. When we got to this village it was snowing like mad. We took the two captains inside a hut and worked on them and worked on them. Unfortunately, Captain Washbrook had died. Captain Farrar-Hockley we did manage to get back. My sergeant-major, myself and Marine Bamfield and a Captain H.J. Pike of the Gloucesters had helped as well. Anyway, the march continued and we reached another camp, Camp No. 3 I think, where the other ranks were shed; senior NCOs and officers went further north still to the officers' camp.

I stayed there for some time, and . . . it was American Thanksgiving. There were just six Brits in the camp then. The rest were Americans, and curiously enough the Chinese organised a special

[1] He had served in the Second World War in the Gloucestershire Regiment and the Parachute Regiment. In 1950 he was adjutant of the 1st Battalion, the Gloucestershire Regiment. Later he became General Sir Anthony Farrar-Hockley, Colonel Commandant of the Parachute Regiment and of Land Forces in Northern Ireland. Author of two volumes of the official history of the Korean War, published by HMSO in 1990 and 1995.

meal for Thanksgiving Day. It's odd, isn't it? They beat you senseless one minute, give you a damned good meal the next, and then starve you for three months. We were already very thin, and it was in this camp that I developed beriberi. It was very much like scurvy where your teeth tend to wobble about in your gums and you develop sores all over your body.

When I reported to the doctor, they gave me a packet of crystals with the instructions to gargle three times a day, which made me laugh. There weren't enough crystals to wash a lettuce. An interpreter outside said: 'Why are you laughing?' I told him . . . what I needed was vitamins. He said we all knew that, but there weren't any. However, some doctors did come from China and saw those who had developed this disease; they shoved needles into us which burned dreadfully. It cleared it up and I never suffered from it again.

We had a bath now and again, at a communal bath down the road which was good during the winter because it was warm. In the summer we used to bathe in a nearby stream, but we had to get permission for that because they weren't too keen on having a load of naked prisoners of war running around. One day I was called to the presence and told I was a reactionary and that I would be going to another place. It was the camp where all the bad boys were sent – and a real bunch of hooligans we were, too. We baited the Chinese guards until they brought in some regular soldiers, and although we had a lot of fun with them we didn't go so far as we did with the People's Volunteers.

There was a clique of us they would interrogate regularly, assuming when there was trouble we would have some connection with it. We got to know most of their tricks and wouldn't give way. Sometimes they got angry and threw us in the cooler; other times we were just sent back to the huts with a warning. I always kept cigarettes and matches sewn in my jacket because I never knew when I might be thrown inside. Actually, we had a wonderful time at this camp, comparatively speaking, of course. Didn't have to do any work, and in the winter they came in one day and asked us if we would like to go on a journey, a voyage down a wide river that ran by the camp. They had an old Korean with an old boat and we used to go down the river and load it high with wood for the winter. We thought the wood was for us, as indeed it was. We weren't doing anything to keep them going. It seemed a miserable way of thinking, but you've got to realise they kept trying to teach us about Communism and to get us to write peace letters, and we just wouldn't do it. Of course, they couldn't see why we wouldn't tell lies for now and say that we agreed with them. So we had to be very careful what we did so that we couldn't be seen to be helping the enemy.

After a while I ceased to look on them as enemy; some of them were so nice, did their job, obviously. Others were just plain pigs, but you get them anywhere. The whole thing was so stupid. We weren't prisoners of war as I understood the German camps had been. It was occasionally like the Japanese camps . . . stuck in a hole for a while for being uncooperative and attempting to escape. I still have marks on my wrists to this day from the handcuffs being pulled tight. I had refused to write letters of peace, and when these Communist journalists, two from Britain[2] and the other Australian, came around, I led a bunch of people who threw stones at them.

Gradually, time wore on and one day the marines in the camp decided to take over the cookhouse because we were getting short rations from the guys from the regiment then handling the food. They were getting the lion's share. I won't mention which regiment they came from; suffice to say that one of them said he was a lieutenant in the IRA. Whether he was or not I have no idea, but he was Irish. I'd just come out of the clink and we marines got together and decided to do a raid on the galley. We went in early one morning, turfed all the others out and we started to do the cooking. We had quite a dust-up that morning, and I was arrested once more for chasing a chap down the road with a butcher's knife, right to the camp gate. The sentry was startled beyond measure and let fly with a round, so I stopped. I was stuck in the hole again and for once I told the truth and they let me go straight away. I told them we weren't satisfied with what was going on in the galley.

Then we discovered that in the ranks there was a guy from the Army Catering Corps who wasn't going to get involved with the other mob but was happy to help the marines. So we started to get decent meals then. Everyone got plenty to eat then; God knows where the rations had gone in the past. The Chinese gave us some steamers, and life became reasonable. We made a basketball pitch and built our own recreational centre. There were a couple of chaps from the Gloucesters who were really good carpenters, and one of them made a clock out of bully-beef tins and it worked. We had a couple of PTIs so we got pretty fit. Made our own weights and what have you. Then we formed our own concert party to keep our spirits up. One Christmas – I think it was 1952 – the Chinese gave us a

[2] Employed by the then *Daily Worker*.

little red book with a peace dove embossed on it. While I was in solitary confinement I used to write in it. It's not a diary but a tremendous jumble of things, including playlets that were performed by the concert party. Then the Chinese came across with a couple of guitars, would you believe?

One day some of us were called to the headquarters, and the troublemakers were all locked up. That was the day we heard the war was over and the Red Cross were coming. We had two or three really good meals and plenty of cigarettes, rice cakes and God knows what. When we came out, the Red Cross had already been and left satchels for everyone containing toothpaste and so on. Next day the Chinese came round with even bigger satchels with the same kind of stuff in. Sometime later, we were all paraded, taken to the railhead in trucks, then on to P'anmunjŏm, where we were exchanged.

There was no such thing as counselling when the Korean prisoners of war returned. Many were seriously ill and received medical care for months; some were to die soon after their return. Those who recovered physically, like Peskett, were returned to their unit, but they were undoubtedly troubled by their haunting memories. Peskett married a nurse he met soon after his release, and she helped him back to good mental health, although at times doing so became a struggle. It got to the stage where he wouldn't go out during the day, only after dark, and he wouldn't wash or shave. 'I was a scruffy devil. The marines just let me alone until one day an officer called me in to see him. He just said, "Talk to me" and I just burst into tears. He said, "Go on. Dig in." I cried for about 20 minutes. And at the end of it, he said, "Now, go home, get yourself washed and I'll see you on parade in the morning", which is what I did. Two months later I was wearing sergeants' stripes. They gave me time – and that's all I needed.'

John Peskett remained in the marines and later gave many lectures and talks on resistance to interrogation. He had much to convey from his personal experiences of the first examples of psychological warfare ever experienced by British troops on such a scale. It became widely known as brainwashing and was identified as a method of influencing people to change their beliefs and accept as true what they previously had considered false. The term was first used to refer to methods of influence practised by the Chinese Communists in the 'thought reform' programmes they developed after taking control of China in 1949.

Chapter Fifteen

The Defector

From documentary evidence and oral statements from prisoners of war from UN countries, it became evident that the Chinese and North Koreans used a variety of techniques to try to convert prisoners to Communism. In the early days, naive procedures were clearly set out in documents originating from the Chinese high command which ordered that 'all captured officers and men . . . shall attend classes and pay full attention to study and make notes . . . All squads should systematically read the issued reading materials . . . Bad behaviours, making noises, joking, dozing are strictly forbidden.'

Later studies showed that the procedures went far beyond compulsory lectures and reading. Prison camps were monitored carefully and prisoners were given classifications ranging from 'unresponsive' to 'progressive' and finally to 'cooperative'. Other forms of indoctrination included isolating victims in a prison cell or a small room, subjecting them to beatings and torture, and telling them repeatedly that their beliefs were wrong. This was followed by starvation and lack of sleep until the torment caused the victims to abandon their own beliefs and accept those of the persecutors. Most victims were found to regain their original beliefs soon after returning to their own environment.

Andrew Melvin Condron became the only man in the British military force in Korea who chose not to come home. He went to China with 22 American defectors. He insists that he was not brainwashed and that what he did was of his own accord, although some psychologists might be inclined to debate the issue with him. His recollections provide, at least, an alternative view, although also confirm the appalling conditions the UN prisoners suffered.

Condron came from West Lothian, from a working-class background. His father had been in both wars and at Dunkirk. He left school at 14 and eventually wanted to join the merchant navy. In the meantime, he took a wireless course at college in Edinburgh and on completion began looking for work. The war had just ended and there was a glut of radio technicians on the market. He was 17, eager for adventure and travel and decided he wanted to become a Commando. He joined the Royal Marines, signing on, as was compulsory then, for 12 years. His parents weren't very pleased because he had studied to join the merchant navy, but eventually they gave their permission. He joined in August 1946 and spent two years in the Mediterranean fleet aboard HMS *Liverpool*. He was sent back to do an advanced signals course, on completion of which he was on draft to join 42 or 45 Commando, who were in Malaya; in fact he was already on embarkation leave in June 1950 when the Korean War broke out.

> A notice went up asking for volunteers for a special Commando force for the Far East. We all knew it was Korea. I seriously considered volunteering, but a friend in Kuala Lumpur had written to me to say there was a good radio station there that he felt I should try to get into. So, in fact, I didn't volunteer. However, a few days later another notice went up stating that most of the places had been filled for the Far East but that there was a lack of specialised personnel – signals and engineers and the like – and so the undermentioned ranks had been posted to the unit. My name was there. In all fairness, it was said that all those who had not volunteered would be interviewed. There were quite a few who for personal or family reasons didn't want to join that unit, and as far as I know anyone who objected was taken off the list. When my turn came to be interviewed, I saw a major and in the end I told him I had no objections. He eventually asked, 'Are you with us?' and I said, 'Yes, sir'.
>
> I think most of us thought the war would soon be over. I had no real thoughts of the ramifications politically. I came from a working-class background and had never grown up with any real political convictions; politicians to me at that time were just professional people doing a job. I suppose on the whole I was anti-Communist but in a very simple way. I'd never taken a great interest in it. There was just this general feeling among people of my age that Communists were trying to take over the world. Another factor in it was that my brother had been in the RAF and taken prisoner at Singapore, captured by the Japanese and worked on the Siam railway. He was a prisoner for three years. He survived, just. And I remember chatting to him when he came home. He

didn't like to talk about it too much but occasionally he would relate stories: that they were maltreated and beaten up and, although the Allied soldiers were harshly treated, the Japanese treated their own soldiers in much the same way. I suppose he had a much more objective view of it than many other accounts that I had read. He didn't like them; he hated them, but he did have a viewpoint and mentioned, incidentally, that the Koreans in the Japanese army were more savage and sadistic than the Japs.

So I already had a dislike for the Koreans when I went to the Far East, and these thoughts were in my mind more than a serious political viewpoint. We flew out to Japan, from the Commando training school, all in civilian clothes. All was supposed to be hush-hush and secret. We came into London in our civilian clothes but carrying our kitbags with 'RM Commando' on them. We travelled out in two groups, flying from London airport; we passed through Cairo and if anyone asked any questions here we were to say we were a football team. We went to an American army camp, Camp McGill, and we set out in our summer uniforms, jungle-green, trained in the use of American weapons and landing off submarines and troop carriers.

We then began doing raids, still based in Japan. I was on the USS *Bass*. She would lie off the coast about five miles. They would disgorge our rubber boats at one or two in the morning, hook them on to a motorboat and be towed two or three miles offshore, then we would row in for the raid. There were three or four boats containing six or eight men. We were lightly armed, the heaviest being Browning automatic rifles. They were well-planned raids and we were well briefed by marines officers in conjunction with the Americans. They were mostly targets that were important and impossible to hit from the air.

We usually sent in one or two men, swimming ashore to reconnoitre, and they would signal us to come ashore. We would beach the boats, blow up tunnels, railway bridges and so on and make our escape before the explosions went off. On this type of raid, if we met opposition, we shot to kill. We came in eight men in a boat and we went out eight men in boat; there was no room to take prisoners unless you were on a specific mission to do so. We did encounter civilians on one occasion on the beach as we were making our escape, and we just tied them up so that they could free themselves later.

We continued doing this for a while, up to about October 1950. The Commando was then sent to the mainland of Korea, by which time the US and South Korean forces had pushed up towards the north and it looked as if the war would soon be over. The whole Commando then joined up with the 1st Marine Division. They weren't quite sure if the Chinese had come into the war, and if they had by what numbers. There were many rumours. I was given to understand that we were supposed to get in behind the lines carrying small arms and be supplied by air and try to make some intelligent estimate as to what sort of numbers. It was an intelligence-gathering operation in some respect.

We were in a convoy heading towards the Changjin Reservoir on the north-east coast of Korea. There was also an element of the US Cavalry, the 7th Regiment, there. We had 30 or so trucks moving up with a few tanks upfront. There was a lot of snow around and the convoy was attacked as we moved along a very narrow road in a valley. At first we came under intermittent fire which gradually became heavier. The Koreans and the Chinese were dug in around the hills and we were sitting targets. They mortared trucks ahead and trucks behind, effectively blocking the route forward and back. We were blocked in and they then gradually closed in on the central column. Quite a few of our people managed to get through; the last 20 trucks were caught in the trap. I was among them. We were ordered off the trucks and dived into a ditch on the side of the road. You couldn't always see the enemy on the hillside; they were in foxholes, and it was a very confused and chaotic situation. We were under heavy fire and returning it. The battle lasted 17 hours, until well after dark. We had only a Royal Marines officer with us, and we concluded we were pretty well stuck. I tried several radios but couldn't raise anyone at all.

The officer came around and asked for volunteers to try to break out. I could see the situation and decided to have a crack at getting out. About 12 of us volunteered; we crawled and half-ran along the ditch and made our way down to the east end of the road and came to an open area we had to cross – 200 yards at least. The ground was covered in snow, some drifting up to six feet. There were trucks blazing, mortars coming in, flares going off, many casualties around. We now had to get across this open ground and decided to do it one at a time, with ten seconds in between. The first six or eight got over and then I was among the last three, and when it came our turn, the Chinese had spotted us. We started to run across and they hurled everything at us. It was like being in a bees' nest, with bullets flying all around. We threw ourselves flat and then we would make a dash again, not necessarily in a straight line.

We were actually stuck in this place for about an hour, and when we eventually got to the other side we discovered the others had already left. We set off in single file, following a stream. A little further on, there was a shot, a single rifle shot, and the NCO who was leading us fell back. He was dead when we got to him. Then we heard voices and dived for cover, but soon realised they were Americans. They had just seen this shape, this figure, coming forward, and the Americans understandably shot him. He was wearing a green beret, not an American steel helmet. I shouted out and called angrily that they had shot my mate. I warily came out and was met by an American captain and two sergeants; they had an American unit dug in by the stream. I asked the American captain if he had seen any of our lads, and they hadn't. By now, the firing had died down and they suggested we stayed with them until light. My clothes were frozen solid now, because we had waded waist-high in the water.

As dawn broke people started moving around. I went down a slope and came across one of our drivers who had been hit in the hip and had crawled to safety. I was kneeling down, tending his wound with my rifle slung over my shoulder, when I heard a noise and turned around and saw what I thought was a South Korean soldier standing there, and just ignored him. He made a noise – Uuh! – which sounded like some sort of instruction, and then prodded me with his rifle. I stood up about to land him one when one of the Americans shouted out from above: 'Hey, buddy, you'd better throw down that rifle. We've surrendered.' It turned out that we were totally surrounded, and I called back: 'Well, thanks very much. Nobody tells me bugger all!'

I slung my rifle down. I was so peeved and pissed off that no one had told me they had surrendered. This guy I thought was a South Korean was in fact Chinese and for some reason he ran over and shook my hand, probably misconstruing my action in flinging my rifle away. Eventually, a Chinese officer came up, gathered us all together and gave us a little speech. There were about 30 or 40 of us in this group. He said we were not his real enemies; the real enemies were the imperialists who had sent us to kill women and children. He said we were proletarians and so was he and their argument was not with us. I wasn't quite sure what a proletarian was, nor did many of us. One guy next to me said: 'Proly-what? I thought they were bloody Communists.'

This speech lasted about ten minutes, then he wandered away and we were marched up the hillside and ushered into a little hut. We'd been in there about an hour or so when one of the guards came to the door carrying a very large gourd filled with steaming hot water which he put inside the door. We proceeded to wash up in this. When we'd finished, the water was soapy and muddy. The guard came back, looked at the water and went berserk. He jumped into the hut, swishing his bayonet around, kicked the gourd of water away and stormed out. A little time later, the officer came back and explained that this guard at considerable risk to himself had built a fire and boiled the water up to drink. We had washed in it and he saw it as a great insult. No one had explained to us that the Chinese only drank boiled water. We tried to explain, but he wouldn't listen. We had made the guard lose face; it was East meets West. Total misunderstanding and ignorance on both sides. Although I didn't see it at that time, it was to me an evident lesson of the difference between us.

People who were badly wounded were loaded into Jeeps and taken back to our own lines. People who were walking wounded stayed. Later on, of course, they died, because we had no medical treatment whatsoever. At the time we had no idea there would be no medical attention. From there we were joined by other prisoners, now about two hundred in all, and we started off on a long march, walking for about three weeks through very, very deep snow on unbeaten tracks up and down the mountains, occasionally strafed from the air, and so we started only to move at night.

The Americans and the Australians had air supremacy, and the pilots had orders to hit anything that moved. I supposed we covered five to ten miles each night through fairly thick snow. The wounded found it very difficult. We carried ours as they gave up and eventually, as we got a little further, they provided a couple of ox carts. But the fact remained that there was no medical attention whatsoever, and I cannot recall any one of the wounded among us surviving that march.

We were guarded by a unit at company strength; they had no transport themselves and carried their own gear, including very large cooking pots, like big black woks. I was impressed by the fact that they didn't make us carry anything. We occasionally stopped off in villages where we stayed in village houses, 20 men to a room. The Chinese would send a man on ahead to negotiate our accommodation. The curious thing was that the Koreans usually refused payment from the Chinese. It was something that affected my thinking later on. We next came to a town which had been bombed and strafed by American planes. Before we got into the town, the

Chinese guards brought us together and made closed formations around us, and we proceeded through what was a very hostile crowd, because they had been hit by the American planes.

We finally got to a village somewhere in the mountains, and we were distributed among the various mud huts – mud floors, straw on the floor, no water or whatever. It was called Camp 10, but wasn't really a camp as such. We had to stay inside the huts all day and were allowed out at night under guard for exercise. For the rest of the time we all just used to sit around in these huts, huddled together for warmth. We couldn't wash, apart from rubbing your hands together in the snow; we were all unshaven and never took our clothes off. Within a very short space of time we were all suffering from lice; our clothes became covered with lice eggs and all kinds of other bugs. We all had crabs and used to run competitions for the number of lice we could kill. It wasn't till we moved to another camp sometime later that we could get deloused, but it is surprising what the human body will put up with.

We had very little food – two sugar bowls of meal a day. We had a limited amount of water collected from a central kitchen. Food and water were shared very democratically, almost to the last grain by both us and our guards. Most of us who were around 12 or 13 stone to begin with were soon down to 6 or 7 stone. We marched on and eventually reached Camp 5, where some of the men were taken away for interrogation. The remaining prisoners went on strike, and I was among the instigators. We refused to answer roll call and so on. Troops were called and fired shots over our heads. But the men were eventually returned and the strike was ended.

The Chinese were very annoyed. The camp commander felt he had been slighted, lost face. So we had to lose face for him to be appeased. Then they made some concessions. They allowed us British to be formed into a separate company and we were able to elect our own camp committee. I was twice elected as chairman. We even had elections as to who ran the cookhouse. That was important, to make sure that you didn't have someone in there who was taking too much food for himself and his friends. So it was all very democratic. It was curious really. We learned to approach things from their point of view rather than ours, and that way we might get things done. In this way the food situation, for example, improved. For the first six months it was drastic and very bad for most of the first year. People were dying, sometimes as many as 20 a day, mostly from malnutrition. The people who had wounds had died earlier. Others died from what we termed give-upitis – they sat down and gave up. They thought we'd all die, anyway, from hunger. That mostly affected the young Americans.

In Camp 5 we began to get ourselves organised. We managed to get ourselves deloused. The camp was next to a river where we could wash and swim and stay reasonably clean. We tidied the camp up and built latrines. We even got rocks and made paths and managed to get whitewash and paint the rocks white so that at night we could find our way if we were caught short. We also kept the camp clean. In the early days, because everybody had dysentery and people very often got caught short, it was just left and people were trampling all over the mess. We managed to get it organised and we became a sort of model company. They tended in the British company to be older men, many with wartime experience, and the younger men had generally been well trained, and we were war children who were used to deprivation. Although we never suffered anything like we suffered in the PoW camps, we were probably better equipped mentally than most of the Americans.

At the same time we gave our captors a pretty hard time. We called them names, mimicked them and so on. I think we all felt we were superior to them, from peasant-based economies. Again it was the question of East and West, totally different logic. They did a lot of silly things, such as attempting to brainwash us in the early days of the war. They didn't see it as we did; some of them had been fighting since the early 1930s, first the Japanese and then the Chinese nationalist army. There had been many battles which we hadn't heard about from which large numbers of prisoners were taken.

Many of them were just peasants who were dragooned into the army against their will, and when the Communists took them prisoner they saw how they operated, and installed land reform and so on. To put things into a nutshell, the Communists would capture thousands of these young peasants, subject them to a quick form of brainwashing or ideological reform for a month or two, at the end of which many of them came out convinced that the Communists were right and joined the Red Army. So in the Chinese Communist experience, it worked. The Chinese Communist is an animal who believed in what they were and were doing. Many were educated people who saw this as the only way out for China. They didn't see it as most people in the West, where Communism became a dirty word. So all in all the Chinese Communists had a great deal of success in ideological reform.

When it came to Korea, they saw us as another form of their

peasantry. They hadn't a clue as to what life was like in the West. They thought it was simply a matter of converting us to their way of thinking, although not, of course, by recruiting us into their army. In the early days, in Camp 10, the indoctrination followed the line of little speeches that they had no quarrel with us, and most of these attempts at brainwashing failed miserably because they were approaching it from the premise that they used on the peasants. The disorientation and psychological techniques were not used against anyone I knew of, but I have read that it was used elsewhere.

On the whole I see it this way: if the Chinese take us prisoners they want to convince us of their way of seeing the world through Marxist eyes, and if they try to force it on us by ill-treatment we were not going to accept it. It would be counterproductive. If they capture someone who was vociferous, they may have used methods of torture or beaten him up or what have you. I cannot deny that. But anything I say was from my own personal experience. My own view was that generally the Chinese were not going to try to impose it on us by force, simply because it would not make a lot of sense. They tried it through their ignorance of life in the West; they did not understand the sophistication of Western life. They weren't stupid, but naive and unsophisticated in their view of us, just as we were in relation to their life and views.

In the early days we had lectures about once a week, traipsing down a mile or two to a big barn, and sat around while this bloke spouted about imperialism and internationalism and so on. We didn't want to listen to this guy haranguing us, which incidentally was all done through an interpreter. We just weren't interested. All we were interested in was getting something to eat and keeping warm. We may have talked about it among ourselves later, but their attempts to brainwash us in this heavy-handed way was a total failure. They stopped trying around the end of 1951. Later on, there were attempts to indoctrinate us – but sometimes it was achieved through simple agreement. I came later on to accept many of the Marxist tenets but not because I was told to, simply through reading. Again, you have to watch periods in this recollection. This occurred after we had organised ourselves and were getting food and were keeping ourselves clean; we were by then above starvation level.

Once you reached that physical stage, you might become more interested in what was happening around you. I became more interested in what they had to say. There were pamphlets for us to read, some produced by the Communist Party in China, some by left-wing organisations from America. Some of us began to take an interest, not so much from the Chinese point of view but in Marxism. I had never heard of *Das Kapital* until then. I read it three times there. It made me think. Two other books, *On Practice* and *On Contradiction* written by Mao Tse-tung, had a great effect on me. It was very difficult to understand in the beginning, but after the third or fourth reading it began to make sense. He used Chinese examples to explain Marxism, so you had to learn a little bit about Chinese social conditions. They were answering questions which I'd had either consciously or subconsciously all of my life. As an example, when we were doing the raids in Japan, people were sometimes surprised to discover that ordinary marines *think*, had an idea.

Later on, other books came in. The journalist Alan Winnington [from the *Daily Worker* in London] came in to visit us and I asked him to get us books. He was instrumental in getting a lot of books sent in from China – Russian and British classics, most of Dickens, American writers and so on. The first time Winnington came to the camp, I didn't like him much. He struck me as being a very arrogant person. He had a Chinese uniform on and a fur hat. When I first spoke to him my impressions were increased. He had the air of being a very superior Englishman; that was my initial impression. I didn't mind too much his being a Communist. People had the right to their own beliefs. I was myself no way committed – I was never in any way committed. Our company invited Winnington to come and give us a talk.

We had bits and pieces of news but we did not necessarily believe what the Chinese told us. We used to get a newspaper called the *Shanghai News* which we called the *Shanghai Liar* because we didn't believe anything we read in it. We invited Winnington to give us a talk on it. A lot of the camp turned up to listen. He told us things, some of which we believed and some we didn't. On the whole it turned out to be true, such as the possibility that the Americans were stalling negotiations for an armistice and to get us free. We didn't believe that, but later on it appeared that there were elements of truth. He also gave us news from home. These things were not particularly important to us, but the role of Winnington was a channel through which we could put pressure on the Chinese to provide things like razors, books, a gramophone and so on. He was instrumental in doing this through his friendship with the Chinese and their acceptance of his judgement. We weren't inundated with stuff, but we got more. At the end of the war, of course, it was claimed he had brainwashed people, interrogated them. I don't know about that; I never saw it. He might have got a bit carried away, or excited. I didn't see any of that.

He took mail from us and some of the first mail we received was after his visit. They always said our mail never got through. I don't think that was true. He wrote the book *Breakfast with Mao*. When he came, living conditions had already improved, so it must have been early 1952. Later on, particularly after I went to China, I got to know him quite well. My original impression of him was moderated somewhat, but he was always a very reasonable man to me, had his own views.

Wilfred Burchett [a writer for French left-wing newspapers] I saw only a couple of time, whereas Winnington came three or four times. He seemed to be a bit more down to earth than Winnington and, being Australian, tended to be a bit more chummy. I got to know him fairly well later on in China. He was a very straight-forward person. I used to have arguments with him. Winnington would argue analytically, whereas Burchett was closer to reality. Winnington could produce arguments that were difficult to handle.

They were both journalists, covering the peace talks. I knew them both, and while I did not agree with their viewpoint or ideas I felt that to accuse them as being traitors or withdrawing their passports was totally unjustified. I know that the Americans felt they were a thorn in the side, giving to the newspapers things that the US military wished to keep quite. They would obtain transcripts from the Chinese side and then pass them on.

My own decision stay on in Korea arose, I think, through . . . my experiences since capture and through the prisoner-of-war camps. It was a harsh experience in the sense that you came very close to death yourself and saw your friends dying around you and you buried them, although you became immune to that eventually. I had developed a political interest in ideas, which I might have had to some extent beforehand. But right up until the last minute I had never had any other idea in my head than to come home. It wasn't until quite late in the day, when the armistice negotiations first foundered and then were agreed, that the thought crossed my mind about staying.

I had been quite impressed by the Chinese at times; they could be silly, but so could we. So I had the idea that it might be interesting to visit China and see for myself, and then perhaps move on to Russia – at most I felt I would stay for a year. There was then the question that I was still in the Royal Marines. It didn't bother me that much. I couldn't see that it was illegal. Under the terms of the armistice agreement, they gave me the option of coming home or staying. It was my decision and I made it. A lot of people, of course, disagreed with that, but it wasn't important to me then. I can readily admit to an element of romanticism in that China had already intrigued me.

The hardest thing to overcome was my family and how they would take it. I came from a close-knit family and I had a large degree of apprehension as to whether they would accept it, and understand what I had gone through and how I had changed. There was nothing I could do about that except hope that eventually I would explain, which in due course I did. It was a decision I made over the course of three days. But then, unexpectedly, the Chinese did not make it easy. A senior officer interviewed me and we went through it. He tried repeatedly to talk me out of it and suggested that the best thing to do was to go home and then return later if I felt so inclined. But I had made up my mind. I was going to stay. There were a good number of other British soldiers at the camp, and other camps, who were thinking about staying, but I believe they were all talked out of it. In the end, I was the only Briton who stayed in China, although there were 22 Americans. I cannot say that they all stayed for my reasons. I did not see myself as a Communist, although I was a Marxist. I was not a convinced Marxist, but a lot of my thinking was along that way.

In the meantime, groups were already leaving the camp and in the end there was only the group of us left – those who were staying – and we remained in the camp until officials from our own side came to interview us to ensure that we were not being coerced into staying or that we had not been brainwashed. We were told that we could change our minds and that no action would be taken against us. Two of the Americans changed their minds and decided to go back. One was a guy named Batchelor who, on his return, was sentenced to 25 years; the other was Dickenson, who got about 10 years.

I think they were the only two who went back from P'anmunjŏm. We stayed there for a couple of weeks, we were given clothes and shoes and put on a train. That was it; we were on our way to China. It was much the same as I had anticipated. The Chinese did nothing to help us in terms of learning the language. I found them very friendly and convivial. Later on, having stayed in China for eight or nine years, I felt I was very fortunate in going at that time, in what I think were the good old days, the best days. On the whole, I learned a lot. I taught English language for a number of years at Peking University. It was quite enjoyable. I grew to have a great love and respect for the

Chinese people. I inevitably ran into bureaucracy; it was and is very much run by bureaucrats, and it was inevitable that you were dealing at the middle level with bureaucrats.

I had quite a number of unpleasant experiences but on the whole nothing drastic. When I was at the university, for example, in around 1956, I had a Chinese girlfriend, a student, although not one of mine. We started to see one another at weekends and had been going out together for four or five months when the family started to put pressure on her. Her father, who had come back from Hong Kong, was anti-foreign. The Chinese are quite xenophobic in this way. The university also got involved and eventually they broke it up. She was told that I was working for British intelligence, which annoyed me because I felt the Chinese should have trusted me. But obviously they didn't. They told her not to say anything to me, that action could be taken against her. She, of course, told me, which tied my hands because I couldn't mention it.

At that stage I felt that if that was the way the university saw me, I no longer wanted to remain there, nor in China. You cannot exist in China without an organisation of some sort. Everyone is part of an organisation, even down to small communities. The organisation under whose auspices I went to China was the Chinese Red Cross. I told them what had happened and told them I was going to leave. They said I couldn't. I made an appointment to see the director of the university and told him what had happened and that because of that I was resigning. He denied it, of course, but then pointed out that no one could resign in China. There was no such word. I said: 'Just watch me. I'm resigning.' I got my bits and pieces together, got on my bike and rode out of the university.

I had not contacted the British embassy. I was classified by the Admiralty as being a deserter because when I decided to stay behind they had been presented with a dilemma. They could say I had been killed in action, missing in action, taken as a prisoner of war or was on the run. They chose the last. But I wasn't a deserter because my status was made clear under the terms of the armistice. So I was still under the Red Cross. They had a little hostel and I stayed there. I told them I was coming back to Britain. I'd had enough. They were a bit concerned about this; there was a question of face involved.

Although I still had a great love for China, I had become anti-bureaucracy, and if I had come back at that time the propaganda might have affected them: they would lose political face. I wasn't particularly worried about it and I was ready to take my chances in

Britain. I felt I had done nothing illegal. They might have considered I was a traitor or let down the Royal Marines. OK, I accept that, but in my mind I had done nothing wrong. I was so irate at the way I had been treated by the Chinese bureaucracy, I decided to move on.

I sold all my possessions and prepared to leave. I met up with one of the Americans who also stayed and he suggested that we went on a bit of a holiday to the Bo Hai gulf. We stayed for a month, and that's where I met my wife. To cut a long story short, we got married and I stayed on in China for a few more years. I told my wife when we got married that I intended to come back at some point. Then my son was born in 1960, and when he was about a year old I decided that we would leave China and come back.

Then bureaucracy came into play against us. The Red Cross said: 'Yes, comrade, you can go, but your wife and child who are Chinese citizens cannot.' So, fortunately by this time, I understood the Chinese mind. I didn't get angry, but decided to take it to a higher level and it took me 18 months to get permission. I employed a number of tactics. I wrote a very long letter to my father; I had it taken it out of the country for me. I asked him to copy it in his own hand and send it to Chou En-lai, the Chinese Prime Minister, which he did. I emphasised two main points: I had my father say he was an old man and hadn't long to live – which was a total lie – and that he wanted to see his grandson and daughter-in-law and his son he hadn't seen for years. And it worked. Chou En-lai apparently read a summary of the letter and said: 'Let him leave. Why stop him?' I made arrangements to leave very hurriedly. I left with two suitcases, a wife, a two-and-a-half-year-old son and £30 in cash We got back here in 1962, stayed in Scotland for about 15 months. It was a very bad winter. I went for a walk one morning and a car drew up alongside me. I was staying at Alloa at the time with my sister. This guy got out, showed me his identification, and he was from Stirling CID. He said that people at the Admiralty were very keen on talking to me, to settle up details about people I had seen killed or buried. So I agreed.

I went into the old War Office in London. There were five men there. One was a psychiatrist; there was a Chinese expert and Foreign Office people, intelligence. I went through the information they wanted and the conversation veered into why I had decided to stay. They were interested to discover that, if I wasn't a madman, they wanted to know from a political and military perspective what made me tick. It ended up I was there for about five days. It was basically how I'd got on, what I'd done, how I felt. Not so much

> intelligence matters. I couldn't have told them as much as their own agents. I spoke of my political views, that I didn't agree with the domino theory of Communism spreading through the Far East. It was just a long discussion about China and the Chinese people and that the Communists were the only people around who could actually do anything for China, and that, if nothing else, history would recognise Communism for making China a nation and freeing Chinese women. I don't know whether I'm right or wrong . . . who's to say?

Condron continued to insist he had not been brainwashed. Others might not agree. In September 1996 US Colonel Phillip Corso was a witness at a Senate hearing into Missing American Military Personnel from both Korean and Vietnam Wars. He recalled that during the Korean War he was head of special projects and intelligence in the Far East command and in due course met and interviewed returning sick and wounded prisoners of war, released from the Communist camp, and received a great deal of information on medical experiments on our prisoners. He told the hearing:

> The most devilish and cunning were the techniques of mind-altering. It was just as deadly as brain surgery, and many PoWs died under such treatment. This was told to me by our own returning PoWs. Many PoWs willed themselves to death. My findings revealed that the Soviets taught their allies, the Chinese Communists and North Koreans, a detailed scientific process aimed at moulding prisoners of war into forms in which they could be exploited. Returned prisoners who underwent the experience reported the experts assigned to mould them were highly trained, efficient and well educated. They were specialists in applying a deadly psychological treatment which often ended in physical torment. Many were simply to be exploited for intelligence purposes and subsequently eliminated.

Chapter Sixteen

Dear Mum and Dad . . .

The domino syndrome of Communism in the Far East that was eventually to explode into the horrors of Vietnam and defeat first the endeavours of the French and then American and Australian troops ran parallel to troubles around the old stamping grounds of the Commandos in the Middle East. Hard on the heels of Korea, the British began pouring troops into the island of Cyprus where EOKO terrorists began their fight for freedom from colonial rule. Colonel Grivas, a former Greek army officer, took his guerrilla army into the hills and began ambushing British soldiers and killing Turkish-Cypriot policemen. Across the Mediterranean, a powder-keg situation was developing around the whole of North Africa in the colonial territories of Britain and France. The scene in the mid-1950s looked grim for the imperial powers as Communist and nationalist elements began to rise up against the constraints of foreign rule.

Not least among the British troubles was volatile President Nasser of Egypt, who had overthrown King Farouk and promptly demanded the withdrawal of the 70,000-strong British army on the Nile. The pullout was completed in June 1956, and six weeks later Nasser took possession of the Suez Canal, wresting it from international control and claiming the revenues for his own country. There was dancing in the streets of Cairo when he announced that the Suez Canal Company, in which Egypt sold its shares to the British in 1875, was being nationalised without compensation. The scene was set for some retaliatory action as the British Prime Minister, Anthony Eden, said he would not to allow a man with Nasser's record to have his 'thumb on our windpipe'.

Britain and France were terrified that vital oil supplies from the Persian Gulf would be severed. The French were also concerned that Nasser was supporting the increasingly violent independence movement in Algeria. Without taking America into their confidence, the two nations began secretly to plan a joint military action to reclaim the canal and depose Nasser, by assassination if necessary.

Throughout August, the British military chiefs went into a huddle with their French counterparts to work out a plan. It would ultimately engage the troops of the whole of 3 Commando Brigade, two battalions of the 16th Parachute Brigade and the French Foreign Legion. While tensions were rising in the Canal Zone, Colonel Grivas took advantage of the diversion and mounted fresh terrorist actions against the British forces. Anthony Eden's deliverance of a new constitution, putting Cyprus on the road to independence, was not sufficient for Grivas, who was still insisting on union with Greece. On 27 September, four newly arrived soldiers were severely injured by a bomb in a toffee tin placed in a rest room. All three commandos of 3 Brigade were, at various times, brought in to Cyprus, and both Commandos and paras were on the island as the Suez crisis developed. Jim Green, whom we met earlier during the Malaya skirmishes, recalled:

> Our time in Cyprus was very reminiscent of our Malaya days in that we were dropped on hillsides and mountaintops looking for EOKA terrorists. The terrain was totally different, of course, but again it was a beautiful country. We were largely engaged on internal security, controlling riots, stop and search and so on. We would also go out into villages suspected of harbouring terrorists, turf them all out of their houses, line them up, and an informer would be sitting in a covered wagon with a little peephole. The villagers would be parading in front and we would pick out the terrorists. We took part in many searches and set up many observation posts. Quite often we would find ammunition and arms and arrest terrorist suspects. They would be identified by the informers, and we would cart them off to the police station. What happened to them after that was not our responsibility. It was lovely country to be in, and for R and R we went swimming in Limasol. Again, we had to be careful of ambushes; several units took quite a few casualties there over a period of several months. But generally there were few close-quarter encounters.

Meanwhile, the British and the French military planners were having trouble finalising a plan for their repossession of the Suez Canal called Operation Musketeer, and in the three months between August and the end of October no fewer than four different sets of operational orders were issued and then cancelled. Fiasco was already a word springing to mind. The final draft was produced in the third week of October, by which time the Israelis had arranged some significant side action by amassing 30,000 troops along its 120-mile border with Egypt, and their own invasion was swift and unhindered.

The Commandos and paras were the key elements in the assault on Port Said, but even the 'final' order of battle was to change when the force was already on its way to Suez, as Major-General R.W. Madoc explained in a summary of events now lodged at the Royal Marines Museum:

> The mission of the Commando Brigade in the final operation orders was to land at Port Said and seize the town, the harbour and the area to the south of the city, firstly with the British sea-borne assault on either side of the casino pier by two commandos, Nos. 40 and 42. H-hour for the landing was to be 6.45 a.m. local time. It would be preceded by air strikes, and naval gunfire would destroy beach defences. The two commandos would set up a beachhead to allow a squadron of Centurion tanks to land behind them. Ten minutes after the seaborne landing, the French seaborne assault on Port Said would begin. Thirty minutes after H-hour, a British parachute battalion was to drop on Gamil airfield, and French parachutists were to drop on Port Said. No. 45 Commando would then land by helicopters north of the interior basin and take the bridges and causeway. Having achieved those objectives, the three commandos would then dominate areas within the city while the parachute troops mopped up any resistance as they advanced forward.

The Commando brigade went aboard a convoy of 20 ships assembled at Malta and, even as they set off on their journey, changes to the operational orders were made. Madoc received a signal from the Allied Force Commander cancelling the helicopter assault by 45 Commando and allotting the task of seizing bridges to the French paras. 'This meant a completely fresh appraisal of the Commando tasks,' said Madoc, 'and the brigade was spread among the 20 ships with the landing only a few days off. I decided 40 Commando would carry out its tasks as before, 42 would advance south to the southern end of the city and 45 would stay in reserve until, when ordered by myself, it would land by helicopters within the beachhead and support 40 and 42 as required.'

However, on 4 November, now two days away from the attack, fresh orders were received. The parachute troops, previously set to go in after the Commandos, would now land a day earlier than the main assault, on 5 November, to effect a surprise attack ahead of the main assault, which President Nasser knew was heading in his direction. The seaborne assault would go ahead as planned, except that no naval guns would be fired except in reply to Egyptian guns, and a ban was placed on the use of guns larger than six inches in calibre. 'To put it mildly,' said Madoc, 'these orders were slightly confusing and up to the last moment of the run-in a large number of individuals were not at all sure whether there was to be gunfire support or not. I was one of them.'

The same confusion was met by the commanders of the paras as they prepared to go in, and at the last minute two battalions had to divert to a seaborne landing. Originally, the entire 16th Parachute Brigade was to have dropped on strategic targets but – as with Arnhem and so many other occasions – Britain didn't possess enough aircraft to lift them in one go. The RAF could supply only enough Hastings and Valettas to carry 668 men, which was to be 3 Para, a miserable 6 Jeeps, only 4 trailers, 6 106-millimetre antitank guns and around 170 supply containers. Thus, just one battalion, 3 Para, landed at Gamil 24 hours before the seaborne assault came in and were met by Egyptian self-propelled artillery pieces and a good deal of machine-gun fire. Several paras were shot as they came down. The biggest and most upsetting pieces were lorry-mounted batteries of rockets which came hurtling over during the landings, and there were many acts of bravery and heroism as the paras, as yet alone in the battle, began to make headway towards their objectives. As the convoy of ships appeared off the coast soon after first light, there were already palls of dense black smoke over the city from burning oil tankers.

Madoc's seaborne assault went in, and the landing was carried out without much incident. The entire force was landed within two hours of arrival, including C Squadron of the 6th Royal Tank Regiment, which joined 40 and 42 Commando in their advances to their specific objectives. They encountered a good deal of small-arms fire and grenades en route and spent most of the morning in house-to-house fighting until, by nightfall, they had achieved their objectives. No. 45, meanwhile, had landed by helicopter into a hail of machine-gun fire in the beachhead area and fanned out in support of the 42, which had become bogged down at a heavily defended warehouse. The commanding officer of 42 had called for air support, but none came. They moved ahead, with 45 in support, and what happened next, recalled by Gordon Burt of 1 Para, highlighted the chaotic state of British preparedness for this expedition:

> We faced the situation of being short of aircraft for striking. I only ever saw two Hunters, and they had to go back to Cyprus to refuel. When you called for an air strike you had to wait ages for anything

> to happen. This manifested itself again when the Commandos went on. The marines called for an air strike, and it didn't come in. They were left there for quite some time, an hour or so, and decided to move on themselves. They had just moved on the objective when the air strike came in, and so that was a sad occasion.

Sad, indeed. What was recorded in the Commando log as 'an uncontrolled British air strike' came in on top of 42 and 45 Commandos, causing 18 casualties, including the commanding officer and the intelligence officer of 45. In spite of everything, the combined forces of the British and French paras and Commandos, with the tanks blasting away at snipers and buildings, pressed on through some stiff opposition, and by darkness most of the given targets had been secured. There were already rumours of a ceasefire, and two senior Egyptian officers had entered negotiations with British force commanders on the ground.

Back in London and Paris, however, the telephone wires were hot with confusion and panic. Indignation and protests were flooding in from around the world, including Russia, naturally, and America. Major-General Madoc had made no headway in talks with local officials, but by 8 p.m. that night he had established his headquarters in a block of flats on the seafront at Port Said: 'It was there that I had my first news of a ceasefire, which came to me over the BBC to which I was listening at the time. An hour later the BBC informed me that the Prime Minister had made a statement to the House of Commons that a ceasefire was being called. At 11.15 p.m. that night I received my orders for the ceasefire to take effect at midnight.'

The Commandos had been fighting in Port Said for exactly 18 hours when they were ordered to call a halt. Urgent meetings were called at the United Nations Security Council. British troops who were within sight of securing the Suez Canal were confused and furious. A poignant letter home from one of them lies in the Royal Marines Museum:

> Dear Mum and Dad,
>
> Last night was bedlam, oil tanks burning, ammunition dumps exploding, bursts of fire and general noise. This morning there was almost a complete change and life in the town came back to normal. What happens next I just don't know. The Egyptians have blocked the canal and of course the pile of shipping is terrific . . . the thing that has struck me most forcibly is that we have got the worst armed and equipped infantry here. The Egyptian weapons of Russian make are first class and the French are better equipped than us. As ever the troops have been magnificent.

And in a second letter from aboard the ss *Empire Fowey*:

> Well, we are out of Egypt and on our way to Malta. This is a splendid ship. The food is excellent and I have done little but eat, sleep and drink. We left all our transport and heavy equipment in Port Said and we remain at short notice to fly back should it be necessary. Somehow I don't think it will be. I just don't understand the politics of this affair or just what made us stop. It must have been either a threat from America or Russia. It will be interesting to know the truth of the matter . . .

The truth was that political wrangling had intervened. There were threats of 'modern and terrible weapons' from the Russians. But worse for Eden was great pressure from President Eisenhower, who was furious and spitting F-words that he had not been kept informed – especially as the attack was launched on the eve of the US presidential elections. Eisenhower, running for a second term, didn't want to send the Americans to war – or even support his former Allies in one – and certainly not on election day. In the event, he won and thereafter showed great hostility towards the British and the French for some time to come.

The invasion caused a run on sterling in the world money markets, and the US would only intervene to help the British Treasury if it were given cast-iron guarantees that the British would pull out. They did, and so did Eden, resigning six weeks later. By 21 November a 6,000-strong UN force moved into the Canal Zone and Britain was out for good. All in all, it was a shoddy ending to a disastrous misadventure.

One other news item: on the same day that the British force landed in Egypt, 1,000 tanks of the Red Army rolled into Hungary and crushed the Hungarian Revolution. Soon after dawn, the last words broadcast by Radio Budapest before it was silenced by Soviet troops were: 'Help . . . Help . . . Help . . .'. Eisenhower couldn't be bothered with them, either. Two days after the UN Security Council voted for an Anglo-French-Israeli withdrawal from Egypt, it told the Soviets to get out of Hungary. The Russians eventually complied – 33 years later.

By now, the last remnants of Empire were fading fast. The Malayan crisis ended in 1957, at least as far as Britain was concerned, with the country's independence. But trouble persisted in Cyprus, and the Commandos were called back immediately after Suez. Colonel Grivas was still giving the troops a run for their money, and the British government finally agreed to seek United Nations mediation for independence, although it would be another two years before a peace deal was hammered out. Much to the chagrin of the military commanders who had been on the ground there, Archbishop Makarios, chief ally of Colonel Grivas, became president. In the intervening two years, the Commandos continued with other elements of the British military to police the island, with EOKA keeping the pressure on almost to the last.

Even as the British troops were pulled out of Cyprus, however, fresh conflicts in both the Far and Middle East would split 3 Commando Brigade virtually for the next decade, first with 45 Commando assigned to the British colony of Aden, followed two years later by the departure of 40 and 42 Commandos for the jungles of Borneo. All units were in for years of hard slog but undoubtedly the heroes of the 1960s decade were the men of 45. They went to Aden and the Radfan in 1960 and remained there for seven long years in the most difficult conditions. They were among the last British troops to be pulled out of the region; they served the longest single tour of any British military unit in the Middle East, and when eventually they returned to England in 1967 they had been in continuous service overseas since 1946.

The tensions in the Middle East followed on from President Nasser's personally proclaimed victory in Egypt against the imperialist forces of the West. In the aftermath, a group of young military officers in Iraq murdered the young King Feisal and his uncle Crown Prince Abdullah, while the Prime Minister, Nuri el-Said, was kicked to death by a mob as he pleaded for calm. The coup immediately put pressure on Jordan, Iraq's partner in the Arab Union, and on 31 July 1958 2,000 British paratroopers landed at Amman while American marines waded ashore at Beirut. As unrest spread throughout the Middle East, British political and economic influence over a string of sheikhdoms, sultanates, monarchies and protectorates became the focus of bitter reaction among Arab republicans.

The one remaining vestige of colonial power in Aden which Britain had ruled for 128 years hung tenuously in the balance. The last strategic base there was a fortress at the mouth of the Red Sea through which southern access was gained to the Suez Canal. This piece of British-controlled turf sat uneasily among a string of tribal sheikhdoms spread around the southern Arabian coastline along the Gulf of Aden which formed the South Arabian Federation. The vast and mountainous lands that stretched far into the distance towards Saudi Arabia became the location of a war between the Soviet-funded People's Republic of South Yemen and the Federation backed by Britain.

By 1960 Aden was surrounded by hostility, and in a chain reaction of events 45 Commando found itself posted to the country to begin what became the longest stint in the history of the Commandos, and where incidentally they would be joined by the newly re-formed SAS. A detachment of the SBS also arrived on the scene, based at Bahrain. They arrived from Malta in March 1960 and took over what was known as the BP camp at Dhala, which became the home and base of the unit throughout its tour of service in South Arabia. Initially, half the unit remained in Dhala and half went to the remote and desolate Radfan region, where they saw action for the first time since leaving Cyprus against so-called Arab dissidents. The Radfan was an almost waterless, baking, mountainous region 50 miles north of Aden. The area was sparsely populated by fierce tribes who ran a tollgate system for caravans passing through their territory on the traditional supply route along the Dhala road up to Yemen. The Commandos had their first taste of internal security duty in August when they were deployed to patrol the fetid alleys of Aden itself at a time when legislation was introduced making strikes illegal. Those first few difficult months set the pattern for the years that were to follow. They became ensnared in the sustained campaign of terrorism and bloody guerrilla warfare of the worst kind across this most inhospitable land. The unit found itself alternating between upcountry operations in the Radfan and internal security duties in Aden, while still committed to carrying out exercises to retain amphibious and limited war skills.

To add to their woes, the newly created republican state of Iraq began eyeing up the oil-rich state of Kuwait over which it claimed ownership. Britain's ending of its protectorate arrangement with the tiny country spurred them into action. In June 1961 intelligence sources reported that the Iraqi army was being mobilised for an invasion, just as Saddam Hussein did 30 years later. The commando was flown en masse to Kuwait in July. There they were joined by 42 Commando, who arrived by sea from Singapore, and No. 6 SBS, from Malta. The SBS detachment performed beach reconnaissance in the event of a major force landing becoming necessary. In the event the arrival of 1,300 or so British marines patrolling on the streets of Kuwait was enough to deter the Iraqis, and in due course 45 returned to Dhala, where its patrols were intensified. Skirmishes with guerrilla forces were becoming more frequent. Indeed, the whole region seemed to be heading towards a total breakdown, and in November the troops marched the 80 miles back to Aden for another spell of internal security duties. While they were in the city, the military staged another show of force for the benefit of local dissidents with a major amphibious exercise involving HMS *Bulwark*, the then-new Commando ship, in the Gulf of Aden, in which 45 Commando played a major part. A few months later a second exercise was staged, bringing together Nos. 45, 42 and the headquarters company of 3 Commando Brigade, again centring around *Bulwark*. A third exercise in the Aden area was staged the following year by 3 Commando Brigade, this time with 40 Commando who came by sea from Singapore and the recently re-formed 41 Commando, which sailed out from England. It was the first time since re-forming that 41 Commando had exercised with any units of 3 Commando Brigade.

Soon, however, exercises were overtaken by the sheer weight of patrols. In 1963 the Yemeni civil war reached the Radfan, and once again 45 was called into the region, now to be joined by successive waves of parachute troops from 3 Para and the SAS. The Radfan campaign settled into a long-running affair for which several elements of the British army became embroiled, including the East Anglian Regiment and the 39 Brigade. It was during this period that 45 began to pioneer the use of helicopters in tactical deployment, a development that was to have far-reaching effects in the future. Their use was highlighted in a report Lieutenant-Colonel T. M. P. Stevens to the Commandant General of the Royal Marines in May 1964 describing a recent action in which 12 dissidents were killed:

> From 30 April to 9 May we lived in the clothes we started in (our packs never caught up), had just enough water to drink and slept on bare rock. We carried three water bottles per man and might survive 24 hours or more on this in static work in the shade. On a long and tough night march the heavily laden men – wireless operators and support weapons – drink a lot. We had no cases of heat exhaustion but a lot of exhaustion! The evenings, nights and mornings are pleasant but the heat of the day on bare mountaintops is very unpleasant indeed. We are proud of our beards in the mountains, necessity, not policy, through lack of water and packs never reaching us. We now know enough of the game to live and fight in that kind of country. The most remarkable lesson was the extreme power of a few well-concealed snipers on high ground can have . . . Logistics is the snag and helicopter resupply almost a critical factor.

The colonel added another sentence of praise elsewhere: 'The operations of the SAS are remarkable. They lie up in small parties among the dissidents, bringing down artillery fire and air strikes whenever they see a target. They must have inflicted 40 or more casualties.'

The commando had a brief respite from the Radfan campaigns and the Aden patrols when they were sent at short notice to Tanganyika to quell a mutiny in the country's army. On their return, 45 Commando served continuously in South Arabia, though the cycle of upcountry and Aden duties became more intense as the situation worsened. The length of time spent on the mountains became progressively longer as the helicopter supply situation improved. In a seven-week period alone, 45 Commando conducted 305 night patrols and, in the early weeks of 1966, mounted 8 major operations. During its last upcountry excursions that year, they logged 24 confirmed kills and after those long treks into the Radfan became the first British troops to leave the area, leaving it mainly in the hands of the paras and 39 Brigade.

The departure of 45 Commando from this wild country did not, however, mean the men would enjoy a period of rest and relaxation – far from it. The British were finally on the way out. Under an agreement formulated by Harold Wilson's Labour government, Aden was being given independence. The deadline for complete withdrawal had been set for 1967, and 45 Commando now moved to Aden and took over internal security duties in the Ma'alla area, which in fact turned out to be one of the liveliest spots during a period of extreme civil unrest to send the British on their way. In one of their last major incidents in Aden, 45 Commando recorded thirteen terrorists killed at a cost to themselves of one killed and two wounded. In the event, they were the last unit to leave Aden, covering the backs of the withdrawing troops being flown out of Khorsmaksar airport on the final days of the British occupation of Aden. During its time there, 45 had collected numerous medals and awards, including three MBEs, two MCs, four MMs, six Mentions in Dispatches, three Queen's Commendations and fifteen Commander-in-Chief's Commendations for operations. All this had been at the cost, to date, of only six killed and sixty-two wounded during seven years in action, which the company log recorded as 'a small number in relation to the time and effort spent on operations'.

And while their colleagues in 45 had been away in the awfulness of South Arabia, the rest of 3 Commando Brigade had been, for much of the time, engaged in another equally difficult and tedious campaign being fought in conditions so very, very different from the heat of the mountains and deserts of the Radfan. This time, it was in the wet and impenetrable jungles of Borneo, where the war that wasn't – it was merely a 'confrontation' – flared when the beady eyes of 'Mad' Doctor Sukarno, President of Indonesia, fell on the British protectorates on the island of Borneo.

At the time Britain was responsible for the defence of the Sultanate of Brunei and the colonies of North Borneo (later Sabah) and Sarawak. Those states, it was hoped, would join the Federation of Malaya to form a powerful and stable alliance. Sukarno had other ideas and backed local Communist guerrillas in a flare-up in Brunei which rapidly gathered pace across the island. The three British protectorates shared borders with Kalimantan, the Indonesian region of Borneo which accounted for three-quarters of its land surface. Sukarno was intent on taking control of the remainder, to add 7.5 million inhabitants to the 100 million he already ruled. His further ambition, inspired apparently by the Japanese in the Second World War, was to take over the whole of the Malayan states and with it the plumb target of Singapore. In 1961 he was poised to continue and the British government approved a hurried, if limited, military response and sent in an initial force drawn from the Royal Marine Commandos, the Gurkhas and the Queen's Own Highlanders and, later, the Parachute Regiment.

Before long there would be 28,000 men deployed, topped up with Australians, New Zealanders and what proved to be a very active contingent of the Special Forces. The latter consisted of the SBS and 22 SAS – one of whose squadrons was commanded by Major Peter de la Billiere (later British commander in the Gulf War) – and elements of the Parachute Regiment who went in to join patrols devised by the SAS. They were all under the overall command of General Sir Walter Walker, who apart from being the Gurkhas' Brigade major-general was appointed director of military operations in Borneo, he being the foremost military expert on jungle warfare in the British army.

Media reports wrote of him as an 'enigmatic' and 'eccentric' figure, which he largely put down to sources at the Ministry of Defence whose noses he had put out of joint on occasions, making no secret of his indifference to some of their edicts; to use his own language, he was under sufferance from fart-arses who, in due course, took their revenge by delaying his knighthood for several years.

Walker was the principal architect of what would become a most successful campaign, with very low casualty figures and innovative in its use of the Commandos and the Special Forces. He did have, however, one hand tied behind his back because the British troops were governed by strict terms of engagements, specific and limiting. They were drawn up to comply with both United Nations requirements and the British government's fear of engaging its forces in a long and costly jungle war in a terrain that was naturally hostile and, in parts, still uncharted. The overall penetration depth allowed in cross-border attacks was just 5,000 yards, although this could be increased to 20,000 yards for specific operations authorised by Walker himself: 'It was a ridiculous situation when you think about it – how can you expect a patrol hacking its way through dense jungle to actually measure 5,000?

They might be 9,000 yards . . . or even more.'

They were known as the Claret Operations. In theory there could be no cross-border operations for the purpose of retribution, and the MoD was insistent that absolutely no member of the British forces should be taken prisoner and absolutely no British casualties, dead or alive, should be left behind in case they might be photographed and produced as evidence to the world of British attacks.

No. 42 Commando was first into the fray, opening its account on 10 December 1962 in a campaign that would detain them for the next three and a half years. The Commando's L Company went first by air to Brunei, where the British resident, his wife and seven others were being held hostage at Limbang, 12 miles south of Brunei town. It was estimated that 200 rebels were surrounding the resident's house. The company, with a machine-gun section, went in at 6.15 on the morning of 12 December and in a furious firefight released the hostages unharmed, although five Commandos were killed and five wounded in the action. The enemy lost 35 killed and at least that many again wounded.

Thereafter, 42 had a highly successful start to their campaign, and within a month had captured 421 enemy – the most rounded up by any unit in the time. They were hampered, however, by heavy rains in January which caused major floods across their patrol area, and Lieutenant A.D.M. Morris and his troop sailed away when the rising water level lifted the complete house they were resting in. That mishap apart, 42 set in motion an operation to find and capture one of the most senior enemy figures in their area, Yassin Effendi. Lieutenant Colin George and Sergeant Smith formed a special team to track him, and eventually caught and arrested him in March.

Thereafter, 42 settled down to a continuous programme of amphibious and jungle patrols in liaison with the rest of the British forces. Their incidents logs ranges through many skirmishes and ambushes and includes another first – the use of a hovercraft on river patrols.

No. 40 Commando arrived on 14 December 1962 and they were to remain in action in Borneo for just three months short of four years. The initial deployment included Pugforce, for A Company, run by the redoubtable ex-SBS commander and Second World War veteran, Major P. G. 'Pug' Davies. They went straight into patrols and raids and then, in April 1965, the unit began the innovation of operating a curfew across a belt five miles wide from the Indonesian border while night ambushes were laid. The unit itself was spread over a front of 140 miles, manned by 32 radio stations in continuous contact. Strongpoints were set up all along the front, and any one of them could be reinforced by helicopter within ten minutes. Later, 40 Commando was the only one to add its own airborne troop when it took responsibility for an area 140 miles long and 140 miles wide at its deepest point.

For newcomers to the Commandos during this period, it was often an eye-opening experience, in more ways than one. Major-General A. M. Keeling recalled his Borneo story:

> As a young and gullible second lieutenant I found myself in 1963 as a troop commander in 42 Commando at the tender age of 19 with Brunei campaign veterans in the jungles of Sarawak. We didn't then realise that we were embarking on what was to develop into a protracted, large-scale operation; we did know that we were looking very hard for terrorists, armed and inspired by the Indonesians, and that they were probably looking for us. That part of Sarawak was still very primitive then and along the Indonesian border, deep in the jungle, there were very few signs of encroachment by the twentieth century.
>
> Our lot was to patrol the border by day and to ambush crossing points by night, and to do this we lived a very simple life alongside the Dyaks in their villages. In that sort of situation, where the privileges of rank almost completely disappear, you all get to know each other extremely well. This, of course, is part of the fascination of this type of soldiering. We also came to be almost totally assimilated into village life, for we shared all the privations of the Dyaks' meagre existence. We came to accept their habits and lifestyle as if we had grown up among them, although in many respects it was a far cry from our own previous experience. We even became quite blasé about the fact that the women dressed only in a sarong and that sarongs in Sarawak are tied at the waist. This, remember, was long before the days of topless bathing. One day, when about our business on patrol, we came across a jungle river with a simple log as a crossing place. This was not unusual. The first two or three men crossed the log to give the rest of us cover as we went over. Just as I was about to attempt to totter over this obstacle without overbalancing into the fast-flowing water beneath, a young woman, dressed as previously described, emerged from the jungle on the far side of the stream and started to make her way across the log towards me. She was a very pretty girl and I can only assume that the look on my face made it transparently clear that it was not only the fact that I was waiting for her to cross that had rooted me to the spot. Quick as a flash the marine behind me said in a rather unnecessarily loud whisper, and mimicking a parade ground order: 'Take a pace

forward, sir, to catch up with your eyeballs!' Royal, in his inimitable fashion, had captured the moment perfectly.

Other than the humour that the Commandos forced into their plight, the 1960s was a busy but in many ways unsatisfactory time, given that the opportunities for amphibious work had diminished substantially, especially in the case of 45 Commando. They had suffered the continuous grind of attrition in Aden and the Radfan and managed to keep up their amphibian skills only by the deliberate action of training. The other two units did at least see water, and plenty of it. Nor was the decade of indifferent postings over. As 3 Commando came back to their bases in England, there were other stirrings, closer to home, that would also very quickly involve them. What became commonly known as 'The Troubles' erupted on the streets of Northern Ireland out of civil rights demonstrations in 1968.

On the backs of those, the IRA returned to its traditions of bomb and bullet and, as history has well recorded, 'ere long 20,000 British troops, like the Commandos, fresh from the colonial wars, would find themselves patrolling the streets of the province. No. 41 Commando was the first to be deployed there, in September 1969, and had the honour of being the first to be fired on while patrolling in Belfast. They were followed by 45 Commando, whose own baptism of fire came when they placed 400 of their men in between two opposing groups, of 2,000 Catholics and 3, 000 Protestants, in Ardoyne.

Throughout the 1970s, new recruits of a youthful age found themselves in the line; not a year passed without 3 Commando Brigade being represented by one or occasionally two of its units, as in the launch of Operation Motorman in 1972. Both 40 and 42 joined a massive sweep through areas of Londonderry and Belfast to bulldoze the barriers set up in the infamous 'no-go' zones from which law and order had been prohibited. Thereafter, and for years hence, it was patrol after patrol and countless special operations across the province, and especially in the bandit territory of South Armagh. Circumstances beyond their control pushed the Commandos further and further towards the role of infantrymen. They didn't like it. But they had no choice.

Chapter Seventeen

Going South

Royal Marine Commandos had a better reason than most for wanting to get down to the Falklands. After the invasion of the islands by Argentinian troops on 2 April 1982, pictures had been splashed across the front pages and the television screens around the world showing British marines spread-eagled face down on the street in Port Stanley inside a semicircle of 'victorious Argentine Special Forces'. It was a humiliating image that angered the British public and the military, not to mention Prime Minister Margaret Thatcher. Those British troops were among the small Falklands garrison under the command of Aden veteran Major Mike Norman, a bear of a man with long service and still only 38 years old. 'And I can tell you this,' said Mike. 'The lads would not have been lying there if we had not been ordered to surrender by the governor, Rex Hunt. In doing so, I reckon he saved our lives.'

Well aware of the small number of British troops stationed in the islands, the Argentinians sent over an invasion force consisting of a marine infantry battalion of 680 men led in by a company of the Buzo Tactico, their equivalent to the British SBS. At the time, the Falklands garrison was unusually large – 80 men in total because the detachments were just changing over. The new one under Mike Norman had recently arrived from England to replace the existing detachment under Major Gary Noott. Even so, the 80 men were somewhat dispersed: 3 officers and 66 men of the garrison manning various positions, plus 2 officers and 9 men aboard the Royal Navy faithful old South Atlantic patrol vessel, HMS *Endurance*. In the event of a fight, Mike Norman could enlist the aid of the local defence force – all 23 of them – plus a former marine corporal, Jim Fairfield, who had met and married a local girl and had moved to Port Stanley. Put up against the invading troops, they would be outnumbered at least ten to one on a man-to-man basis, but outgunned by a far greater margin. Even so, Mike Norman's lads staged a spirited resistance to the Argentinians before they were forced to surrender. The Buzo Tactico went straight to the marines' barracks at Moody Brook, which was empty, but they trashed it anyhow, tossing phosphorous grenades in most of the rooms. They then moved on to Government House, where the bulk of the marines had assembled to protect the governor and his small staff.

By then, further elements of the invasion force were coming ashore, with armoured troop carriers and some heavy weapons. Lieutenant Bill Trollope's defiant group of 8 Royal Marines were holed up at a point where the convoy would pass, and Marine Brown claimed a hit on one of the armoured vehicles with a 66-millimetre rocket which must have upset the 50 or so men inside. The group followed up with heavy bursts of machine-gun fire, and the Argentinians were temporarily halted. There were a number of similar firefights going on before the inevitable happened. Similarly, the Royal Marines garrison at Grytiken on the island of South Georgia, consisting of 20 marines commanded by Lieutenant Keith Mills, gave the landing parties a few shocks, first bringing down a Puma helicopter and then damaging a landing ship with their 82-millimetre Carl Gustav antitank rocket, the largest gun they possessed.

When the Argentine navy started shelling them with its 100-millimetre gun, the Royal Marines contingent had no alternative but to give up; there was nowhere on that inhospitable landscape to hide. Back on Port Stanley it was also decision time. Mike Norman reflected on that moment years later when he had retired from the marines and was enjoying a more mundane life as a self-employed gardener:

> Surrender, as far as we were concerned, was never an option. The men felt the same. But Rex Hunt, the governor of the Falkland Islands, ordered me to do so, and he was Commander-in-Chief. I was devastated. My men knew the score. Although we were overwhelmed, they still wanted to fight on. I was terribly proud of them for that. The fact remains, we were not there to defend the Falkland Islands; that was impossible for such a small group. We were there to defend the seat of government and that was Rex Hunt. I gave him three options and none of them involved surrender. I told him we could carry on fighting, we could do a runner with him into the countryside and wage a guerrilla war, or he could negotiate a truce to give us some breathing space. He told me: 'Mike, I want you to order your men to lay down their arms.' That decision undoubtedly saved our lives. I have to say, however, that the invasion was no surprise to us or to Rex Hunt. The captain of the

> *Endurance* had been giving warnings for a year which were ignored in London. Mrs Thatcher and her government could have done better. They got us into that situation, and I told her so after I had a few drinks at a Falklands dinner a few years later.

Later that day, Rex Hunt and Mike Norman were told they were being repatriated, along with their men, to England. They were herded off the island, quietly promising that they fully intended to return, and in due course they could, as Juliet Troop attached to 42 Commando. In the meantime, arrangements were made to send an RAF VC-10 to Montevideo, from where they were flown to Brize Norton immediately. There, apart from a large media presence, was a whole range of military and intelligence people waiting to debrief them. What was needed most was information – because by then Mrs Thatcher had wanted a task force to get down there pretty damned quick and reclaim the islands.

There had, as Denis Healey put it so succinctly, been a major cock-up; about that there was no doubt. The Thatcher government, less than three years in office, was engaged in heavy cost-cutting, and Defence Secretary John Nott took a machete to military spending. The Royal Navy's sole bearer of the White Ensign in the southern hemisphere, the ice-patrol vessel HMS *Endurance*, was to be withdrawn in the summer, thereby saving a miserable £3 million a year in a budget counted in billions. General Galtieri, head of the Argentine military junta with problems of his own, decided that a bit of national fervour would do no harm, and sent his fine young men to take the Malvinas. It was something he had obviously been thinking about for some time, as the skipper of *Endurance* had been warning all and sundry in London for months.

Britain being the classic absentee landlord in regard to the Falklands, few in Whitehall had actually been to the place and, furthermore, had no idea as to how or when they could get them back. The islands were so far away. Another couple of centimetres on the map and they'd be falling off the edge of the world! Mrs Thatcher decreed immediate action. Thus, the mandarins and the unfortunate Foreign Office and Ministry of Defence people into whose hands this hot potato was dropped were more concerned with the when rather than the how. For them speed was of the essence, and, as everyone with a modicum of knowledge about military planning knew full well, hasty decisions cost lives. In one remarkable top-level meeting between the suits and the uniforms, according to one of the latter who was there, 'some bloody fool (among the former) suggested making a very large bang in the area of Argentina'. For a while chaos reigned in the corridors of power until the Supreme

Commander herself took charge and announced a task force of 40 ships would be assembled to carry two brigades and a large assortment of top-quality supporting artillery, intelligence and reconnaissance units, not to mention the brilliant naval logistics and management teams.

Although it has often been written that virtually nothing was known about the coastlines and landing possibilities around the Falklands – which is why Mike Norman and co. were so heavily debriefed on their return – this was not at all true. In fact, the Commandos had their own walking encyclopaedia, Major Ewen Southby-Tailyour, who had commanded the Falklands garrison in 1977, had conducted an exhaustive study of the coastline and carried out the only detailed survey of many of the beaches and their approaches since the mid-eighteenth century. Barely three years after his return he found himself at the forefront of the planning team being drawn together by Brigadier Julian Thompson, commanding 3 Commando Brigade, for the move south.

It would take five weeks for the main convoy to reach the Falklands but even as the troops began to embark the task force ships on 6 April, it became apparent that Mrs Thatcher had called for an early response to Galtieri well in advance of the arrival of her boys. The target selected was the island of South Georgia, 800 miles north-east of the Falklands and not actually part of them. It had a small Argentine garrison but also possessed the most inhospitable landscape, with whirlwinds whipping up driving snow across treacherous glaciers. Unlike the army, all Commando units had regular tough winter training in Norway, and 42 Commando had not long earlier returned from theirs. It was thus an obvious choice for the main force to be sent on ahead to capture South Georgia.

It was commanded by Major Guy Sheridan, second in command of 42 Commando, a very experienced snow and ice mountaineer who had only recently completed the first traverse of the Himalayas on skis. M Company, 42 Commando, commanded by Captain Nunn, was selected for the task, and he took with him two eighty-one-millimetre mortars, a section of the Commando Reconnaissance Troop, a small logistic and medical party, two Gunfire Observer parties from 148 Battery, one SBS Section and an SAS Squadron: 230 men in total. They flew to Ascension Island on 7 April, embarked the huge Royal Fleet Auxiliary tanker *Tidespring* and sailed south under the protection of the destroyer HMS *Antrim* and the ageing frigate HMS *Plymouth*. Two nuclear submarines shadowed the group, which reached the island on 21 April.

The SAS were first ashore, landing reconnaissance patrols on the hazardous Fortuna glacier against the advice of Sheridan. Within 12 hours they called for helicopters to evacuate them immediately because of appalling conditions and impossible terrain. Two Wessex Mk 5 helicopters from the *Tidespring* were sent to extract them but both crashed on the glacier when the pilots became disorientated and lost the horizon which merged with the sky in 'white-out' conditions. An ancient single-engine Wessex 3 from *Antrim* piloted by Lieutenant-Commander Stanley in suicidal conditions brought the SAS team and the crashed helicopter crews back to the ship, leaving the force's only two troop-lift helicopters destroyed on the glacier.

Then the *Tidespring*, with the main body of M Company on board, set out to sea after reports of a prowling enemy submarine in the area. The Argentine submarine *Santa Fe* was sighted on the surface five miles out from Grytiken by Lieuentant-Commander Stanley. He attacked her with machine-gun fire and depth charges, which stopped her from diving until two British missile-carrying helicopters arrived to deliver a more devastating blow. The sub was left badly damaged and listing heavily.

Although *Tidespring* was at that point miles away from him, Major Sheridan was determined to make the landing on South Georgia. He mustered a scratch force of 75 men from M Company headquarters, the Reconnaissance Section, the SBS and SAS aboard *Antrim*. He was given permission to lead the group into attack, and with the backing of guns from the destroyer he headed for the target area on South Georgia and before long the Argentinians ran up the white flag. The garrison consisted of 137 men who, on 26 April, became the first prisoners of war. Back in London, Mrs Thatcher dashed out of the front door of No. 10 Downing Street and in front of television cameras shouted: 'Rejoice, rejoice. South Georgia has been recaptured.'

Meanwhile, the task force was heading south after a stopover at Ascension Island to gather more troops and ships coming in from Gibraltar. Ahead of their arrival, SAS and SBS patrols were going ashore on the Falklands proper, inserted on 1 May to carry out the reconnaissance of enemy positions. With them were naval gunnery control officers who were to direct naval gunfire on Argentine positions. Some of those units spent up to three weeks ashore, roaming across the windswept landscape gleaning intelligence. The advance shipping battle group was already giving the Argentine positions a 'bloody good pounding' as the task force sailed to within sight of the islands on 16 May. The mini-D-Day was set for 20 May, when the bulk of the troops would go ashore, although there was a good deal of cross-decking and changing units about from one ship to another before the final moment of the landings.

SBS and SAS units would be positioned on the beaches to guide the troops in as they were ferried over by landing craft from the assortment of carrier ships that had brought them down, ranging from North Sea car ferries to the sedate liner *Canberra*. Although the navy group considered that it would have air superiority, the commander of 3 Commando Brigade disagreed, and he was right. Fortunately, many troop landings were conducted in foul weather, with dark, low clouds and rolling banks of mist to cover the movement of around 5,000 men eventually going ashore. The 3 Commando group – the only brigade with ancillary support groups in the British military trained in amphibious warfare and Arctic action – consisted of 40, 42 and 45 Commandos, the 29 Commando Regiment, Royal Artillery and sappers from 59 Commando, Royal Engineers. Others in the group included elements from the Logistics Regiment, an air squadron running Gazelle and Scout helicopters, an Air Defence troop, the 1st Raiding Squadron operating assault craft and the Mountain and Arctic Warfare Cadre. The 2nd and 3rd battalions of the Parachute Regiment bolstered the infantry strength, and there were two troops of light tanks from the Blues and Royals and a battery of the Air Defence Regiment with Rapier anti-aircraft missiles. Following behind, to arrive a few days later, was the army's 5 Brigade with the Scots Guards, the Welsh Guards and the 7th Gurkhas.

Ahead lay terrific battles, largely undertaken by the paras and the Commandos: 2 Para was the only unit given two battle missions, first taking Goose Green and then Mount Longdon. In the former, storming into the attack against the well-dug-in positions of 700 Argentine infantry on Goose Green, they lost their commanding officer. Lieutenant-Colonel H. Jones, a tough and courageous officer with a reputation for leading from the front, was killed when he personally broke into the Argentine defences fearing that his attack might stall. He was mortally wounded in the process and was posthumously awarded the Victoria Cross.

Then there were the great yomps – 'march or die' – across the entire length of the island, notably by 3 Para and 45 Commando to their targets in the mountainous ranges and hills towards Port Stanley as the enemy were pushed further and further east. The battle plans accounted for all units with the exception of 40 Commando, who dejectedly learned they were to be held in reserve. In fact, they were found constant employment and two of their companies were used to reinforce the Welsh Guards, who suffered a hundred casualties, including fifty-six killed, when the carrier ships *Sir Galahad* and *Sir Tristram* were attacked by Skyhawks.

In those eastern ranges, ferocious fighting under fearsome shelling from both sides saw many acts of daring and courage and singular examples of bravery: 3 Para had a major fight to clear Mount Longdon, and the selfless heroism of Sergeant Ian McKay saved the casualty figures from being much higher when he single-handedly charged an enemy position, hurling grenades to relieve 4 and 5 Platoons who were pinned down and in danger of being wiped out. The sergeant's attack was so close that 'he dispatched the enemy but he was killed at his moment of victory, his body fell forward into the bunker of the enemy who had killed him'. He, too, was awarded a posthumous VC. 'None of us,' said General Hew Pike, who was at the time a lieutenant-colonel commanding 3 Para, 'had ever before experienced what we were going through that night. Mount Longdon, like all the other battles of the Falklands, was close-quarter infantry combat. And that makes for a very nasty, brutal, dangerous, exhausting and scarring experience.'

Similar encounters were met by 2 Para, in their second battle assignment to capture Wireless Ridge, while 42 Commando took Mount Harriet and 45 Commando attacked Two Sisters. The army's 5 Brigade was given three other mountain positions, although only the Scots Guards came to blows. The Welsh were late in starting, which also delayed the Gurkhas, and by then the Argentinians were flooding off the mountains to surrender against the firepower of the British troops and their artillery and naval batteries. The battles leading up to the eventual surrender have been well documented. For a more personal viewpoint, and private thoughts in the events leading up to the surrender, the diary composed by Captain Ian Gardiner, commander of X Company of 45 Commando, makes fascinating if disturbing reading as he explores the sights and sounds of moving towards battle. These edited extracts follow his unit's progress from the moment it went ashore at Ajax Bay from the carrier ship:

> 20/21 May: Two hours late because the Paras ran into trouble, we climbed over the guardrail down the scrambling net. Further delays when one of the landing craft broke down somewhere. Back alongside to cram the remaining people into three landing craft. Had there been any opposition at all, we would have taken a hammering. We secured our objectives and deployed to the hill. Nearing the ridge I saw my first enemy aircraft. He whipped along the ridge line from north to south, very close to where I was. James Kelly, 1 Troop Commander, and I hit the deck. In a split second he was gone and was shot down soon afterwards.
>
> Our Rapier anti-aircraft missile detachment had by now arrived (late), but it was not in a position to deal with the four Mirages that had eluded the Harriers. One was shot down by a ship but the other three pressed home their attack on HMS *Ardent*. I saw these aircraft as they whooshed over us and over our hill to strike *Ardent* in the

Falklands Sound. Our Rapier fired one missile but the missile failed to respond to the controls and missed badly. *Ardent* had not sunk but limped out of sight. We consolidated our positions and dug like little moles. That night, we froze. Our bergens [backpacks] didn't turn up.[1]

22 May: Early in the morning, I sent Alastair Cameron, my artillery observer officer, to set up an observation post at the top of the hill. On arrival he gave me a chilling description of *Ardent* burning and eventually sinking stern first. The shock of actually being at war was beginning to sink in. I was to see things in the next few days that I thought I would never live to see, and hope I never see again. The serious bombing started that day. Several times throughout the day, Skyhawks appeared and bombed and tried to bomb the large number of ships in San Carlos water. I still cannot understand why so many valuable ships were crammed into that small stretch of water. There was no need for them to be there. They couldn't manoeuvre and the land screened their radar. They couldn't fire their weapons at will because they had to be careful not to hit us and the numerous helicopters ferrying stuff around the place. Indeed, I saw several very near misses on helicopters and our own troops by friendly guns. We couldn't fire at will because of the danger of hitting them. They could have remained at sea and come in to unload at night when required. I believe the navy generally underestimated the air threat for much of the war.

That evening, two frigates limped in. *Argonaut*, a Leander, came in being steered by two or three landing craft. She evidently had sustained damage to her steering mechanism but otherwise looked pretty healthy. *Antelope* came in under her own steam a little later. She was making smoke, rather dirty wispy smoke, unusual for a type-21 frigate and had a hole either side of her hull forward of the funnel. The radio mast on her funnel was bent over, apparently having been struck by a low-flying aircraft. Otherwise she seemed sound enough. We were completely unprepared, therefore, for the huge explosion that erupted from her soon after sunset. I was actually watching her as she went. A large lump of superstructure flew through the air as the explosion ripped through her. From the magnitude of the explosion, it became evident early on that there was no saving her. This had obviously been clear to the captain as

[1] They should have been delivered by helicopter. Commando units are traditionally resupplied by air every 24 hours when on the move. In the Falklands, that became a virtual impossibility through weather and shortage of helicopters.

the crew began to gather on the flight deck. By now, although the sun was down, all was bright as day. Apart from the light of the fire, a number of helicopters had very quickly taken off from HMS *Fearless* or *Intrepid* and illuminated the rescue with their underslung lights. Much bravery was done that night. A number of secondary explosions were taking place and yet helos [helicopters] continued to fly. I could see the crew quite clearly in their orange immersion suits through my binoculars.

There was no question of panic and all appeared very orderly as landing craft came alongside, completely disregarding their own safety, which was by no means assured. Soon, almost all the crew were off. The ship aft of the funnel was well ablaze. Suddenly about a dozen men appeared aft of the bridge. One brave nameless landing craft coxswain came alongside and took them off from a ladder. No one even got their feet wet. It was a most impressive rescue and all spontaneous as far as I can gather. Only three died – all men who were near the unexploded bomb, for that was what it was – as 'it' exploded. My deepest respect goes to those remarkable men who fiddle with bombs. They are the bravest of all.

I watched fascinated while *Antelope* burned throughout the night. How can something made of metal be so combustible? I had been in her in Hamburg and there she was, a British ship, dying before my eyes. I can understand why sailors, and particularly captains, get emotional about their ships. She carried on burning and exploding throughout the night until soon after first light when her back broke and she folded in the middle and sank like a tired old lady sitting down. She went quite slowly in a cloud of steam.

23 May: The Ajax Bay Airshow continued: for the first time, ground forces were attacked. Out of a wave of four Skyhawks, two attacked ships and two headed for the beachhead. Both land-bound aircraft dropped bombs. I was particularly concerned because half of my company had gone down the hill to eat a centralised slap-up meal of stew and bread. There were no casualties – and the meal was excellent. Again waves of Mirage and Skyhawks came in. One Mirage struck by a Rapier missile immediately trailed heavy smoke and smashed into the side of a hill. Parts of smoking burning aircraft were spread over a wide area. I saw no parachute. A cheer went up from the troops on the hillside that would have done credit to Hampden Park. But it was nothing to the roar of approval as a Skyhawk, coming in very low over the water, and in full view of us on the hill, was struck by something, I know not what, flipped over on its back immediately and plunged into the water with an

almighty splash. We saw the pilot's parachute coming down and watched him being rescued by a small raiding craft.

This was a gladiatorial contest of the highest order which we were watching. Those pilots were extremely brave and very skilful. They could have won the war for the enemy if it had not been for the outstanding performance of the Harriers, which stopped many actually getting as far as the islands, and the amazing Sea Wolf missile. The men of the navy certainly covered themselves with glory on those days. This was the most dramatic day I saw, although other planes came in and some had considerable success. HMS *Coventry* and *Atlantic Conveyor* were soon to be added to the list of losses and eventually *Sir Galahad* and *Sir Tristram*, too.

26 May: We received our orders for the breakout east from the beachhead. We were to march from Port San Carlos, having been ferried there by landing craft, and capture Douglas Settlement some 20 miles away. No helicopters were available. We were to carry our bergens. We moved out before first light the following morning down to the beachhead. A certain amount of kit had been ditched but each man was carrying on average something in the region of 55 kilogrammes. Some were carrying a good deal more. We boarded landing craft and sailed the 40-minute trip to Port San Carlos and started walking. On the voyage, I produced my mouth organ and played a few tunes. This seems to have been appreciated.

The walk from Port San Carlos to Newhouse, some 12 miles, was the worst of my life. The weather was not too bad but the ground was boggy. Where it was not boggy, there were strong lumps and tufts of grass on which one stands a good chance of turning one's ankle. In places it was pretty steep, but all faded into insignificance compared to this cursed weight, much of which I knew to be wholly unnecessary. The marines were magnificent. We lost the first man after 180 metres and about 6 more over the next few hours.[2] They were mostly the weaker-spirited men who dropped out, although they possibly did have something wrong with them. The rest went on with the greatest of stoicism and good humour all day and through until two o'clock the following morning. For those at the tail end of a queue of 650 men bumping and stumbling through the black night, life must have been hell. By the time we leaguered up, I was near my wit's end.

Earlier on that evening, as we stopped for our supper before last

[2] In fact, only 15 of the 650 marine Commandos dropped out; nearly all rejoined 45 Commando within a week.

light, we heard the explosions of bombs in the area from whence we had come. The inevitable was happening at Ajax Bay and the enemy air force was making a concerted attack on the stores, dumps and buildings there. We feared for our people. For much of the early night, the sky was lit with flashes from secondary explosions as ammunition exploded and fires burned. Five of our rear echelon were killed and sixteen wounded. Most of them were cooks. At 2 a.m. I gave the order to bed down without erecting bivouacs. Our bivouacs (or bivvies) were simply waterproof groundsheets, each supported by a small stick. This was a bad mistake. It rained and the plastic bags in which our sleeping-bags were stretched did not keep the water out. Our sleeping-bags were soaked. I awoke around five that morning to find my bag and I were soaked through. After mentally weeping for an hour or so one got used to it.

We refined our methods of living in this inhospitable place to such a degree that by the end we were like animals and almost preferred it out of doors. We could have gone on for ever. At first light we moved in tactical formation in fighting order only, X Company leading, the remaining seven miles to Douglas Settlement. We took it around 1600 hrs. unopposed. The enemy had fled. We were tired, soaked and hungry, having theoretically finished our rations at breakfast that morning. Having advanced in fighting order, we had not even a wet sleeping-bag to climb into. Helicopters were expected to bring our bergens up, but not for the last time they failed to appear. I feared the night would be worse than the last, but this was not to be so. We were able to sleep two-thirds of the unit in the sheep-shearing shed and, although it was extremely cold, it was dry and out of the wind. The bergens turned up next day and, as often happened, it was a beautiful drying day of which we took full advantage. Many men now started to build up small stockpiles of rations against a repetition of their failure to materialise. However, it is very difficult for a fit 18-year-old who has expended so much energy to stop eating the cardboard packs, let alone the food inside. Some men were constantly hungry.

30 May: Completely restored. We were dry, we had rations, blisters had been patched and it was a beautiful day. It was at Douglas that we had heard of the attack of 2 Para on Goose Green. They clearly had done very well although a commanding officer's job is not to get himself killed. Doubtless he will be fêted for his gallantry. I suspect that his frontal charge on the enemy machine-guns was more an act of desperation than anything else. First tales of Argentinians shooting downed pilots in the water began to filter through. On this march, one of my Royal Artillery party must surely

go on record as the best countryman of the campaign. At one of the 10-minute stops after we had just crossed a stream, he leaped back into it and snatched out a huge brown trout. We could barely believe our eyes. We poached it when we got to Teal. It was quite delicious.

We could see Teal hours before we arrived there – which we eventually did in the dark. There was snow on the ground. We spent a blissful night in a filthy, tick-ridden, beautiful sheep pen. Late that evening, the CO gave orders for a general deployment into firm defensive positions around Teal itself. We were actually some way behind the front line having been preceded by 3 Para and 42 Commando. I believe the most widely learned lesson on this operation was that nothing goes to plan – and although it shouldn't discourage one from actually making a plan, one must be ready to have it changed at any stage by circumstance. The commander who survives and succeeds is the one who can keep the aim in mind and pick up the bits of what is left without fuss and tack something viable together and still be ready to have that shattered. FLEXIBILITY!

2 June: We were to fly by helicopter to a patrol base at Bluff Cove Peak. The weather closed in and flying was not possible. We waited a further day and the weather cleared. No helicopters turned up. We walked to Bluff Cove Peak. It took two days. We arrived as the wind stiffened and it started to rain, but we didn't care because we were now 10 miles from Stanley and there was just no way we were going back. We had yomped all this way, the only unit to walk the whole way . . . and we were proud of it.

4 June: We formed our patrol base at Bluff Cove Peak on the rear slopes of Mount Kent. The attack on Two Sisters was to go in on 11 June. During this period, a number of recce patrols were to be made to see what we were taking on. We had seen very little of the enemy as he had always shrunk away before us. The valley to the west of Bluff Cove Peak was a grubby, boggy place, and our position was split up by a number of stone runs. These extraordinary features were rivers of huge boulders and were very difficult to cross. The stones were covered with a slippery lichen-type moss. They often moved when you stepped on them. Crossing them by day was bad enough. At night, it was hell. It was feared that the enemy might use the nearby Mount Kent as an observation post (OP) once it was vacated by 42 Commando. I was ordered to send a troop up there, and Chris Caroe with 2 Troop was duly dispatched to set up home on the top.

6 June: In normal conditions, it would have been a pleasant walk, an ascension of 300 metres over 2½ miles with a view of Stanley

and the rest of the world at the top. Instead, we had a dreadful scramble in quite the most appalling weather. It poured with rain and blew like hell. On the top, which took two hours to get to, a number of people, including myself, were picked up by the wind and removed several metres and deposited on our backsides. Again, the marines were magnificent – so patient, so tough and so friendly. But after sending a couple of clearance patrols out, it was quite clear that we were achieving nothing that wasn't being done already by Artillery OPs and 2 Troop, except exhausting two troops. So we got permission to come down again.

While we were stumbling around Mount Kent, Chris Fox and a team from his reconnaissance troop were hitting the enemy on the Two Sisters feature. Through the mist from time to time I had seen the occasional shell burst in that direction. Apart from inserting themselves right into the enemy position and discovering much-needed intelligence, Fox's team killed at least 12 of them while fighting their way out with no casualties to themselves, other than a shot through Fox's finger.

9 June: David Stewart, 3 Troop Commander, was given a mission to harass the enemy and inflict casualties. He prepared his patrol well and I never saw an operation so well founded. His men were champing at the bit and I had high hopes for his success. My only fear was the lack of cover presented by the bright moonlight.

As I bade them farewell at one o'clock in the morning, I had a last-minute feeling of horrors. I could see them with the naked eye for 275 metres and with binoculars for up to third of a mile. But the enemy never saw them until they were upon him. They moved to the south of the Two Sisters feature, a rock which was to be my company's objective during the brigade attack. There, with considerable skill, they inserted themselves across open moonlit ground into the bottom of the position. They killed two men on sentry at very close range and in the subsequent firefight killed another five confirmed. They were engaged by at least three machine-guns from different positions and had to fight their way out against an enemy defending ground of his own choosing in superior numbers without supporting artillery. With bullets ripping up the bog around them, they skirmished backwards running like hell in bright moonlight to get to the nearest decent cover, across the river 900 metres away. By the grace of God, no one was hit. By any standard, this was a successful patrol and, as it happened, had considerable bearing on the success of our subsequent operations. But all was not success . . .

10 June: A patrol from Y Company was tasked to go round the eastern end of Two Sisters and find out what was there. A section of mortars was detached from the troop to move forward to where they could support the patrol. The rifle troop commander took the necessary precautions to avoid meeting his own mortar section. But the mortar sergeant got badly lost and the two groups met in a place where the mortars just should not have been. The troop saw mortars first. From the way they were going, it seemed likely that this was [an enemy] fighting patrol on its way to attack our position. However, the sergeant checked by radio that there were no other patrols out. He even checked that the neighbouring unit, 3 Para, had nothing about. He finally spoke to his mortar section and asked them what sort of ground they were on. They replied that they were on high ground and close to their objective. The sergeant looking down at them 180 metres away and this lot were nowhere near mortar's objective. By now they were less than 90 metres away. The sergeant engaged them with his troop. After about a minute, he shouted an order and one of the mortars must have heard him because a shout in English came across: 'We are callsign 52.' The troop stopped firing. Eventually, proper identification was made. Four men died, including the sergeant in command. I listened to the latter part of this fiasco on the radio and to the evacuation of the casualties. It was mentioned in the CO's orders the following day, and the rest of the unit followed his excellent example and the whole dreadful matter was put behind us. The time for mourning our dead was not now.

We were to be prepared to move on and attack Tumbledown Mountain on completion of the Two Sisters assault. X Company was to be the leading company. Before last light, we conducted rehearsals on a nearby crag conveniently similar to our objective. We then relaxed and ate a meal. We moved off, all 150 of us, about an hour after last light. The chaplain, Wynne Jones, came with us. He brought up the rear with extra medical stores. I was sufficiently apprehensive to say to him quietly as we gathered to do our final checks: 'Pray for our souls, vicar.' 'I won't need to,' he said; 'I won't need to.' His presence had a curious effect. X Company did not have many who would have professed allegiance to any formal religion. But to have this man of God in our midst was a wonderful source of comfort. It is not easy to describe one's feelings before one is committed to battle. Fear, certainly, plays a part, but it is not fear of death itself. It is more a sadness about the grief that will follow one's death among one's family.

11 June: The march to the forming-up point from which we were to start our assault was a near nightmare. I had not appreciated how much the man-packing of the Milan[3] would slow us down. Instead of covering the relatively easy ground, which had been recce'd in about three hours, thus giving us an hour to spare, we took six hours. A piece of rock or a small stream that a man in normal fighting order would never have noticed became a major obstacle when carrying his own kit plus a 14-kilogramme-weight round of Milan ammunition. We had something like 40 rounds, and the company was constantly being split. To make matters worse, my route recce team manifestly failed to take us the easiest route. We eventually got there, but not before we had stumbled and cursed our way over rocks and cliffs for half the night. I even managed to lose half the company on two occasions, which meant further energy and time wasted going back to look for them. I lost one man, a key signaller, who became ill in the middle of it all. We had to dispatch him to a nearby artillery battery, and one man knocked himself out falling down a very steep slope. Phil Whitcombe, my second in command, managed to resuscitate him. By the time we reached the forming-up point at least 2 hours late we were 150 very fed-up and tired men, and that was before the real work began.

Ten minutes later the assault began. The worst point of all was the crossing of the open ground. We could see, as we approached, the tracer from a heavy machine-gun arching across towards a neighbouring hill from the top of our objective. While 2 Troop were moving up, I decided to see what the Milan could do and invited them to have a go. It would mean firing over our heads and we were nearer to the target than we were to him, but it would fill the gap before the artillery arrived. The round was fired. It must have passed about nine metres above us, and when it hit, it produced a most satisfactory bang. Soon 2 Troop were ready and Chris Caroe had given his orders, but although mortars had given me one barrage of eight rounds, I could not, through Alastair Cameron, get any artillery at all. Poor Alastair was feeling very useless at not being able to provide at the critical moment. I don't really know what the problem was. All I remember is that when I really wanted the damned stuff, it wasn't available. I asked mortars to repeat their mission and ordered Milan to fire another two rockets.

The rockets came but the mortars then failed me, too. Their base plates had disappeared into the soft ground, and they were only able

[3] Wire-guided antitank missile, particularly useful for bunker-busting.

to support us with one solitary mortar. So we forgot about artillery and mortars and Chris Caroe took his men up the hill regardless. It was an impressive performance. Not only were the enemy now using an antitank weapon, they were lobbing artillery shells over too. Caroe's men picked and clambered their way round up and over the rocks towards the enemy position. It was almost like fighting in a built-up area. Two men cover while one man jumps over, leapfrog all the way. When I saw the ground next day my heart missed a beat. It was much more rugged than I had envisaged, and I would have had serious doubts about their chances, but I would have had no alternative as there was only room for one group at a time.

By now he was involved in a right old ding-dong battle at the top of the hill. They actually got on top but were forced off it by artillery. We then took our only casualty. Lance-Corporal Montgomery and Marine Watson were both some one and a half metres from the shell when it exploded. It picked them up and threw them several metres. Montgomery had broken or dislocated something in the shoulder and didn't know whether it was Tuesday or breakfast-time. Eventually, the leading section flushed the enemy out and we were secure on our objective. My headquarters was still being shot at in a desultory fashion by a half-hearted rifleman on a flank. We ignored him. But the shelling began more seriously now and that was unpleasant. From now until 36 hours later we underwent intermittent shelling, and the intermittent nature of this shelling added greatly to the danger and uncertainty. They were only using two guns, but over the period, some forty or fifty shells landed on our ridge among our positions, and the saddle between us and the rest of the unit became heavily cratered with many more. We received no casualties, but it is a disagreeable business being shelled.

The irony is that if there is not too much noise elsewhere, you can hear the enemy gun go off and then the whistle as the shell comes your way. More often than not one only hears the whistle. If a shell is coming really close, you get about a two-second warning, and if it is going to hit you, you get none at all. We had plenty of opportunity to make something of a study of this as we huddled among the rocks. During a lull, I got out to have a pee. I hadn't finished when I heard the dreaded whistle. The shell hit the ground at the same time as I did . . . 19 paces away. A large lump of rock landed on my back. I reached the top of the feature at first light. When I looked back down the ridge and realised what we had done my spine froze. Potentially, the position was impregnable.

The rest of the commando had had their battles, too. They had secured their objectives after a spirited punch-up in which they lost four dead and fifteen wounded. We buried the enemy dead *in situ*. I myself supervised the burial of an unidentified officer and sergeant. They had both been shot at close range, probably by 3 Troop, on 9 June. We put them in a shallow grave, marked the spot carefully and made a record to submit to the authorities. I said a few words in prayer. We found a prodigious amount of ammunition, weapons, food and equipment on the position. Watching 2 Troop next day going through the camp at the top of the hill was strongly reminiscent of watching a family of tinkers picking over a rubbish tip.

As it turned out it was well that so much was there. Our bergens again failed to turn up and it froze that night. Without the Argentinian blankets and other clothing, we would have been extremely uncomfortable. That evening, as I was giving my instructions for the night, a figure appeared out of the darkness on our position. We were somewhat edgy and were naturally extremely amused to hear a shout out of the gloom in a strong Welsh accent: 'Hello X Ray Company, it's the vicar and I've forgotten the bloody password.'

We were there for two days while the Scots Guards took out Tumbledown Mountain. We were then scheduled to take out Sapper Hill.

14 June: it became apparent that the enemy were crumbling towards Stanley and that a gap was appearing between Tumbledown and Sapper Hill. We were ordered to move at best speed to advance on Sapper Hill. It was a seven-mile gallop but our spirits were high. We took the hill unopposed and were extremely surprised to find the Welsh Guards had turned up before us by helicopter. They would have taken a chopping if any enemy had met them. They were clean, well fed, and I thought they looked soft. It was later we heard of their marching fiasco. They had set off from Port San Carlos and the ground and the weather had been too much for them. They had turned back. We were not surprised. They were left behind to garrison the islands after we left, and I felt sorry for them.

As we were advancing up Sapper, rumours started filtering through about a surrender. I think that it was the inner warmth supplied by that knowledge that kept me going that night. It was the bitterest of the war and the most uncomfortable of my life, and I slept not a wink. It was particularly bitter because although the navy could manage a victory flypast of helicopters, they couldn't find one to bring our bergens up before we froze again!

16 June: We marched into Stanley. It was an emotional hour or so. Naturally, having been in the field for a month, we would hardly have cut much of a dash in the Mall. But we were proud enough and I can think of few prouder experiences than walking into Stanley at the head of my company accompanied by Lance-Bombardier Ingleson carrying our home-made company flag. We were greeted by extremely happy civilians. Then the unbelievable happened. A pub was found open! X Ray Company crammed into the Globe Hotel; I have had happier moments in my life, but they have been damned few. Goodness knows who was buying the beer. I certainly wasn't. Standing there in this smoky, homely shack surrounded by 150 tired, stinking, unshaven, filthy, lightly boozed of my own marines was a moment to treasure for the rest of my life.

Chapter Eighteen

Cutting Edge

The settling back process that once accompanied the aftermath of action had changed significantly. Since the end of the Second World War, the marine Commandos had barely been out of action. The progression has been clearly marked in these chapters, from Palestine, to Malaya, Hong Kong, Korea, Suez, Borneo and the Middle East which, prior to the Falklands, had kept them fully employed through two and a half decades. There ought to be a period of retrenchment and renewal for any military unit after any period of action which was both demanding and injurious. Even those with extensive battle experience cannot fail to be affected by the horrors they have witnessed on the battlefield. But far from becoming less demanding as the British military commitments to its former possessions and protectorates declined, the workload remained at a steadily upward curve on the graph throughout the remainder of the 1980s.

Northern Ireland continued to require an average of 16,000 men on permanent duty within the province, and the Commandos took their place alongside the British army units in rotational duties, usually tours of six months, although some faced a residential posting lasting to two years. For the most part, the marines were non-residential, which meant they were separated from their families for long periods at a time. Elsewhere, the Cold War continued to demand the presence of British troops in their traditional buffer areas of Europe as part of NATO contingency operations against possible threats to the security of member nations. Additional threats arrived with the emergence of volatile hostile nations and terrorist groups, unleashed after the withdrawal of European powers from their former colonial territories, particularly in the Middle East. The spectre of the latter appeared in the early 1970s but expanded to an even greater degree in the remaining decades of the century as unstable regimes began to build their arsenals of modern warfare, which would include vast stocks of chemical and biological weaponry. By the early 1990s, no fewer than 21 nations were identified as possessing weapons of mass destruction, and a number of them were working towards a nuclear capability.

As the cutting edge of the Royal Marines, 3 Commando Brigade had to include many new disciplines within its repertoire from an early stage of these developments as a specialist group at the forefront of NATO's constant state of readiness to meet all possibilities. This required constant involvement in major NATO exercises for which British units in general were well to the fore. Recognised as the premier amphibious group anywhere in Europe, 3 Commando Brigade thus established itself as a crucial element in both national and European defence arrangements.

On the home front, there was also a very substantial widening of the areas of operation for the marines and the SBS. The traditional role of ship protection for which the marines were originally formed remained high on the agenda, especially after several terrorist attacks on passenger ships in the 1970s. Specifically, there were no fewer than three operations in this period when marines and the SBS were engaged in protective measures against threatened attacks on Britain's passenger flagship, the *QEII*. Even greater demands on the skills of the marine Commandos became apparent with the proliferation of oil and gas installations off the coastlines of the United Kingdom. Their remote positions created the need for a permanent amphibious defence structure to protect them and other elements of national infrastructure, such as nuclear plants, power stations and the Channel Tunnel, from potential terrorist action.

Ongoing attacks on the British mainland by the IRA and the continuing rise of political and religious extremist groups merely heightened the necessity for amphibious cover around Britain's coast, with a marine group maintained in a permanent state of readiness, prepared and trained for immediate action. These arrangements, first set in place in the early 1970s, needed to be constantly reviewed to meet both the arrival of new installations and the identification of methods of likely attacks by emerging terrorist groups. The Comacchio Company – so named after the wartime battle in Italy – was permanently on short notice to move with the SBS in the event of any form of protective measure being required. Thereafter, coded plans which could if necessary progressively involve the entire commando brigade were in place to cover any attack situation. This entailed having available a full and regularly updated knowledge of the position and internal layout of sensitive installations. Exercises and constant planning for such eventualities were subject to regular change and review and indeed were routinely tested by mock attacks.

By the early 1990s there were additional potential tasks for the Commandos, and the SBS in particular, with the progression of organised crime into the wholesale movements of drugs, illicit arms and human cargoes across the world by sea. These again required even further specialised training by the commando units as a whole. Thus, while the Royal Marines was an established and integral component in the nation's defence plans, 3 Commando Brigade was increasingly considered to be the elite of the British amphibious forces, and it would not be overstating the position to say that elements of the British army were rather jealous of their standing.

The fall of the Berlin Wall and a virtual end to the Cold War did little to diminish these specialist defensive requirements. The collapse of Communism initially created some euphoria among the politicians as ushering in a new world order in which defence spending could be substantially reduced. Cuts in the whole of the British military structure were foreshadowed in the controversial document entitled *Options for Change*. To this, later, was added the possibility of a reduction of troops in Northern Ireland as the various peace initiatives got underway, although early hopes in that direction were quickly stalled. Similarly, the withdrawal of the iron-fist rule of the Soviet Union and the Balkans created its own problems with civil wars in the former Yugoslavia and turmoil in Russia as it descended into gangster-ridden disarray.

Even so, manning levels in all three services were seen to be capable of some reduction. Many regiments were affected, and the long-serving and immensely loyal Gurkhas, whose position was aggravated by the withdrawal of Britain from Hong Kong, had to accept huge cuts indeed. The overall manpower of the Royal Marines was targeted to reach an establishment of around 7,000 men and women. The Commandos, however, were to remain at a fairly static level – totalling around 5,000, including their support units – and to some extent reverted to the concept originally projected by Winston Churchill in 1940: a well-equipped force, highly trained, highly skilled in the whole spectrum of modern disciplines, ready to move at a moment's notice by sea or land.

In this respect, the position of 3 Commando Brigade was to become unassailable. There remained a very obvious and growing need for their specialist skills in so many areas which, like the paras, could not be duplicated by any other force. In an ever-changing environment, and with a manpower that was turned around and shifted about to keep them fresh, constant training and exercising remained prime occupations of the Commandos, as they had done since their formation. To some extent their skills also provide a clear definition of the areas in which they are most suitably employed, and acting as infantry in a desert war was not necessarily one of them.

Like their forebears in the Second World War, they could fight in the line when necessary, but when it came to the build-up for the Gulf War in 1990 the Commandos were not needed, other than for the specialist tasks assigned to the Special Forces group which included the SBS. Matthew Cawthorne, a young officer in 45 Commando at the time, recalled that his unit was about to begin a six-month tour in Northern Ireland and the build-up to the Gulf War came as they were training for deployment to the well-known bandit country of South Armagh:

> We followed the developments of the war with a certain amount of interest although with some detachment, as we weren't actually involved and there was very little likelihood of us getting called away from an operational tour to go there. The only possibility that we saw would be clearing Kuwait at the street-fighting, fighting in built-up areas type theatre. The reason that we saw no role for us in the desert against the main defence positions of the Iraqis was that because we are amphibious and mobile we tend to carry all our stuff on our backs and one marine cannot carry enough water to sustain him for more than a day. That means resupply every 24 hours, and it wasn't thought that that was sustainable for people carrying all their stores.
>
> There was bitter disappointment on an overt level, and people were saying: 'Oh, my God. For the next 20 years everybody in the army is going to be unbearable because they will be wandering around with a chest full of medals and we will have missed the boat.' There was also a certain element of relief, not because we didn't want to do it but we wouldn't have wanted to go straight from Ireland; that would have been very hard, particularly on the married people or on the families. So I think we made a big thing about being disappointed when it was certain that we weren't going.
>
> The whole thing finished just before we were due to return home, so we were on our six week's post-Ireland leave. Odd things stick in your mind at times of crisis. I remember I had just finished watching *Neighbours* on the box, the wife was at work, I had nothing to do, so I flicked through the channels and the Secretary of State for Defence, Mr Tom King, was making an announcement in the House of Commons. As I switched over to BBC2, he said the fatal words 'and the formation I have decided to deploy is the 3rd Commando Brigade of the Royal Marines,' and that was the first indication I had that we were likely to get sent to Iraq.

The deployment of the commando brigade came in the aftermath of the Gulf War. While the coalition of international troops had been evicting Saddam Hussein's forces from Kuwait, a human tragedy of unimaginable proportions was unfolding in northern Iraq. His forces were switched to attack the Kurdish population in the northern part of the country. Bumbling George Bush, President of the United States, played golf in Miami and refused to commit his troops to getting involved in what he described as a civil war. In fact, it was an ongoing campaign of genocide waged against the Kurds as a whole. Fearing another chemical weapon massacre by the Iraqi leader like the one that killed thousands in 1988, the Kurds began fleeing their homes. They moved across country to the Turkish border, where they were halted, and in the east towards the Iranian borders, where a huge mass of people suddenly appeared. In the space of a few days, two million men, women and children were lodged on the hillsides in appalling conditions with no food, water or shelter.

Within a week, the situation was desperate. Television pictures around the world showed the Kurds massing on the mountains, and the human suffering caused by the cruelty of Saddam Hussein and the dithering of the west, and Bush in particular, who had the power but not the will to resolve the problem before it had even begun. It was left to British Prime Minister John Major to devise a 'safe haven' plan for the Kurds, to force Saddam to allow them to rest above the 36th parallel without fear of being attacked. Major also appealed to his European counterparts and the Americans to join the British in sending troops to help resolve the crisis until a UN force could be established. Bush reluctantly agreed, and then later tried to say it was his idea in the first place.

In Britain, 45 Commando, fresh from helicopter patrols all over South Armagh, were recalled from leave to join the entire 3 Commando Brigade for immediate dispatch to Iraq. While other British contingents of the coalition forces were disengaging from the entrails of war in southern Iraq, they would be dispatched to protect both the Kurds and the humanitarian effort being rushed into place by the international aid agencies from attack by the Iraqis. The Commandos joined troops from other nations under the UN banner to force the Iraqis back behind a 'line drawn in the sand' 20 miles south of the town of Zakho. Any Iraqi aircraft overflying the safe haven area would be shot down and any troops on the ground would be forced back to the 36th parallel. Matthew Cawthorne again:

> We hadn't actually perceived our role as protector of the Kurds. The only possibility that we did envisage was going out to help the withdrawal in the south as soon as the main battles were over. But

> there is no question: they badly needed a force which could deploy quickly and be capable of operating in appalling conditions. On their televisions the world had seen Kurds in freezing, wet conditions, starving, dying of dysentery, fighting over small packets of food. Something had to be done about it, and it was politically expedient to get someone out there fast. We were available, so I think it was partly a political thing and partly because we were good people to send.
>
> There was a great dash for us to get jabs to update us medically, collect our NBC [nuclear, biological and chemical] suits in case Saddam Hussein threw chemical weapons at us and draw specialist stores and ammunition. And since we were to fly out there, the RAF were very particular about the way we packed our gear, so that kind of slowed things down a little. We went by a navy coach from Royal Marines Condor, which is on the east coast of Scotland, to RAF Lucas, from there into an RAF VC-10 which flew us overnight to southern Turkey. There we de-planed and were issued with a small amount of operational ammunition and then put into a mixture of American and Royal Air Force Chinooks which would then fly us to our given positions.

The area of operations for the incoming troops was divided into two, northern and southern. The northern strip was called the JTF Alpha area and covered the area where the refugees were spread in their shanty encampments across the mountains. The southern strip was called the JTF Bravo and covered Kurdish towns and villages from which the populations had evacuated themselves in fear of Iraqi troops and secret police. The priorities were twofold: to get food, water, shelter and medical supplies to the starving multitudes out in the open and, secondly, to create conditions in the towns that would encourage the Kurds to return. The first was a test of resources even to the most experienced international aid agencies; the second part of the equation required delicate handling. Matthew Cawthorne again:

> 45 Commando located themselves around a small schoolhouse in a small hamlet, and we then set up a company base to patrol across the region south towards Dahuk. We were confronted by typical Middle Eastern scenes: dirty, low concrete buildings, a variety of vehicles, some rickety, some smart, driving round the place, and the civilians dressed in long robes, people drawing water, washing clothes and all the rest of it. It was pretty deserted when we got there. We saw a very few people to start with because the population had fled into

the hills. In an area originally of some 30,000–40,000 people, there couldn't have been more than 2,000–3,000 when we first arrived, but they gradually started moving back.

We began with hard patrolling when dealing with the Iraqis but when we were around the Kurdish people it would be more the hearts and minds type of operation and very often you would see our lads patrolling down the street with two or three youngsters hanging off them. It was more of an opportunity to get out, talk to the locals, reassure them and play games with the youngsters than it was to dominate the ground in a military sense. The purpose was to increase the confidence of the Kurds so those who were up in the mountains would be encouraged through the grapevine to come down, knowing they had a secure environment. It was all to do with getting them off the mountains.[1]

We would, on the other hand, patrol hard against the Iraqis, particularly if any troops came into the area. We weren't allowed to call them 'the enemy' as we were by then at peace with Iraq. We couldn't talk about guerrilla fighters or freedom fighters; it had to be armed Kurds. The orders varied. Sometimes it would just be, 'Go out, walk around the town, show your face and if you see any secret police, stop them and give them a hard time.' Specific instructions were given: we would be told to go to a secret police house, knock on the door and ask them when they were leaving. If they were still there on a subsequent patrol, we'd tell them to leave and actually build up the pressure on the Iraqi Forces to move out. On one occasion at Amadiyah [close to the Turkish border] we put them in the back of a four-tonner and drove them down to Dahuk to get rid of them. This, it turned out, was strictly at odds with the high-level coalition policy. The coalition had agreed with the MCC [Military Coordination Committee, which included representatives from both sides] that nobody was going to be evicted and that we would be nice to the Iraqis. So when 45 Commando picked up these people and drove them out, it caused a bit of an upset. The Iraqis complained and the unit got its knuckles rapped, which is a shame because nobody had told even the brigade commander exactly what line the troops should be adopting with the Iraqis who refused to budge.

The local Kurds would identify the secret police, although very often you could spot them. They would be smartly dressed, wearing

[1] Other elements of 3 Commando Brigade had gone to the mountainside to encourage the hordes of refugees to return to their homes and to help with the humanitarian effort.

> American sunglasses, smoking American cigarettes, and they would be driving round in smart cars, very arrogant. In one town we were getting on quite well with the Iraqi army, who were running a checkpoint, when this Bathist [government official] turned up. He looked like something out of *Saturday Night Fever*, in a black shirt and a white jacket and black trousers. He was a very nasty piece of work, and the Iraqi soldiers were clearly frightened of him. Well, he asked us our names, which we didn't tell him, and anything we said he wrote down in a little book and made it obvious that he was part of the government machinery for keeping people down.

As the days wore on, the remaining Iraqi troops – some of whom were clearly unaware of agreements reached in Baghdad for them to pull out of Kurdish areas – were, said Cawthorne, 'given an eviction notice'. In one town nearby, a company of Iraqis with two tanks had steadfastly remained at a checkpoint, where they were harassing any Kurds who came through. Cawthorne's patrol turned up in their truck and had been warned in advance a confrontation might result. So the patrol, with radio contact to headquarters to call for air support if necessary, marched forward in staggered formation 'with a certain amount of trepidation and hearts in mouths'. No firing was opened up, however, and Cawthorne eventually talked to the Iraqi company commander, who claimed no knowledge of the 36th parallel ruling and said he was waiting for orders to move. 'I explained that we would really like to see the back of him. The rest of 45 Commando rolled up about six hours later and then the following day the Iraqis left.'

Only once was Cawthorne's unit forced to open fire, and that was at a palace supposedly used on occasions by Saddam Hussein. It was kept under observation for two weeks and occasionally Iraqi soldiers would pop out and fire their rifles in the direction of the 45 OP from a range of about 600 metres. Then came heavier bursts of fire, and the OP commander ordered his men to return fire. A total of 438 rounds were fired by the OP, and two of the Iraqis were seen to fall. They were dragged inside the palace and there was no further shooting. The Commandos did encounter other problems during their stay. A large number of minefields had been laid off the roads, particularly on likely observation positions that had been occupied by the Iraqis. They were antipersonnel mines of a particularly lethal kind, set off by a tripwire attached to five prongs. When activated, the mine would leap out of the casing and explode at waist height, causing much damage to the victim. Although there were no casualties among the Commandos, at least eight people on the humanitarian mission were killed.

Apart from the protective patrols to encourage the Kurds to return to their villages, the Commandos played a major role in the international efforts to provide food, tents, fuel and medical supplies. The Kurdish leaders, meanwhile, were keen to know what arrangements were being made to curtail the activities of the remaining secret police and how long the protective forces would remain. There were also delicate discussions about the setting up of their own community administrative structures to enable the UN troops to pull out as soon as possible. The latter were already underway. The Americans decreed they wanted all their people out by the end of June, and 45 Commando was put on alert for recovery to the UK; 40 Commando, who had been in the north evacuating Kurds from the mountains down to the hills and the plains, took over the 45 positions before they themselves were eventually withdrawn with the rest of the brigade, leaving the aid agencies and the UN to take over. The eventual outcome, however, was a complete takeover of the region by Saddam Hussein and, in the south of the country, the Shias, who had taken refuge in the marshes, were equally suppressed by the ruthless Iraqi troops. What had been achieved by the British and American troops was to clear the very public spectacle of two million people camping on a freezing mountain.

Among the reasons for the rapid withdrawal of the troops was the looming threat of trouble elsewhere. Even as the Commandos were returning home, crisis was unfolding in Yugoslavia. The collapse of Communism elsewhere had caused continuing threats of civil war in Yugoslavia. The ill-fitting jigsaw of nations held together since the end of the First World War was coming apart and finally burst into a horrendous and brutal series of wars that were to continue for the remainder of the century. Fighting between Croatia and Slovenia in the north-east of the country lit the fuse for an explosion of hostilities across the whole region. British, American and European troops would be engaged first under the banner of a UN force to resolve the murderous battles in Bosnia and then under NATO sponsorship to halt the ethnic cleansing of Kosovo. Like many elements of the three services, the 3 Commando Brigade supplied specialist teams and manpower in support of the allies' effort through the crisis.

Throughout it all, they maintained their routine of exercises, training and deployment to various foreign parts for joint operations. And they continued to provide cover in Northern Ireland, which they had done virtually every year since the troubles began, resulting in marines appearing among the list of 448 troops killed and 5,544 wounded between 1968 and 1998. Indeed, a tour of duty in Northern Ireland has been part of the initiation into the service of many young officers and other ranks gaining their first experience of active patrolling. These notes were written by a junior officer with 40 Commando:

> Arriving in west Belfast with 40 Commando, more than 20 years after the first troops did so, in distinctly different circumstances, one is bombarded by vivid sights, unusual sounds and peculiar fragrances, which typify life in Northern Ireland. While some are merely revisiting for a second or perhaps third time, the vast majority are freshers on whom this operational tour will have a profound and possibly lasting effect. Assisting the Royal Ulster Constabulary in the policing of west Belfast in areas such as the Falls Road, the Turf Lodge and Andersonstown as well as the heavily fortified Sectarian Divide along the Shankill is a hugely complicated and often a frustrating task at all levels. Additionally, trying to explain to our loved ones what we actually do during our six-month tour is equally complex. It is a very reactive experience for our marines since they are really only waiting for something to happen. The remainder of the unit is busily employed supporting them, be it operationally, satisfying voracious appetites, maintaining communications or ensuring that there are sufficient televisions for the few hours of relaxation each man grabs.
>
> It is a 24-hour-a-day existence for everyone, with weeks becoming merely a series of lines through dates on a calendar. We are scattered around three locations, including North Howard Street Mill, affectionately known as the Mill,[2] an architectural relic of a bygone era of industrial prosperity in west Belfast. Were it not for the British army, this crumbling monument to the Troubles would undoubtedly stand empty like so many others around the area, awaiting demolition and subsequent reincarnation. There is a comfortable feel about the Mill and a strong sterile smell hanging in the musty air, a product of daily mopping with industrial detergent. Life in the Mill is not unlike life on board ship. The layout of the cabins and offices along long, straight corridors could easily be mistaken for that of a large vessel. No one has a window in their cabin that they can see out of without a step ladder – the lack of natural light gives the place a sense of timelessness, and the piped heating creates a permanently subtropical atmosphere which is a sharp contrast to the wintry temperatures outdoors.
>
> People and vehicles come and go at all hours of the day and night, and the ambient noise of a humming engine is replaced by

[2] An old factory that was taken over by the Military.

the racket of drills and other machinery being used in the unrelenting maintenance of this ageing building. Venturing out of the gates of the Mill or one of the RUC stations, the reality of west Belfast life presents itself to you immediately. Many of the disastrous urban housing 'solutions' of previous decades, such as Divis Flats, have been replaced by rather desirable brown-brick, low-density housing estates, but the amount of abandoned domestic waste is unsightly and, at first, rather alarming. Moving on and remembering not to judge a book by its cover, life seems almost normal, and in many ways it is. Then a series of heavily armoured grey police Land Rovers roar past you and soldiers clad in combat uniforms, wearing helmets and carrying rifles, walk around the corner amid mothers with prams, and your vision of 'normality' is shattered.

Moving on still further you come across a road with a heavy gate bolted across it severing a major thoroughfare and a wall as high as the house next to it. This is the 'Peace Line', which keeps Catholic and Protestant separate. Walking around west Belfast as a marine, reactions are varied. You are, by virtue of reading accounts of the Troubles, conditioned to expect hostility to your presence in Northern Ireland; the reaction in reality is less hostile. You can expect obscenities from local youths as a matter of course (and peer pressure). Most people simply look straight through. Some will pass the time of day, though little else.

At night, however, away from watching eyes and twitching curtains, people of all ages will talk freely and may even wish you luck. You seize this consolation, for tomorrow again you may not exist in their eyes. The people of west Belfast are fiercely proud of their history and demonstrate their allegiance to the Crown or the Campaign for a United Ireland with huge colourful murals, expertly painted on the ends of the predominant terraced housing. Other walls are daubed with news of the moment, be it support for the peace efforts or vows never to surrender to the South.

Back at the Mill, men are sleeping, manning operations rooms, seeing to blistered feet, working out in the gym or standing by to react to the next incident. Dawn will break again over our small part of this large and varied city and a new day will begin. For some, dawn means the end of a 14-hour shift and a chance to sleep. You may arrive here with a detached and objective perspective on Northern Irish life, but Belfast will surely leave you with lasting memories and perhaps a little perplexed. Memories of the good days, the bad days, the people . . . and of the comradeship between the marines, perhaps, for ever.

The latter phrase has been a recurring thought throughout. The Commandos have moved on vastly from the days of the eccentricities of Lord Lovat, Jack Churchill and others of the old brigades. The Commando spirit remains, and indeed the preamble to the Royal Marines' web site begins with the words: 'The brigade's unique *esprit de corps* is derived from the commando ethos . . . the word originated in the Boer War where a commando was a small self-sustained fighting unit, carrying everything it needed with it and which conducted highly effective raids against the enemy – the British.'

Which is where we came in . . .

EPILOGUE

Modern Times

As we have already noted, the Commandos came under the auspices of the Royal Marines in 1943 and under the overall commander, the Commandant-General, who is based at Whale Island, Portsmouth. The whole of the RM Commando in 2000 was around 7,000 men and women serving in RM establishments and operational facilities at home and abroad.

The combat units are entirely within the remaining 3 Commando Brigade which today has three Commandos: 40 Commando, based in Taunton, 42 Commando, in Plymouth, and 45 Commando, in Arbroath. They are the brigade's equivalent of light infantry battalions and may be deployed independently or as part of an operational group. Individually or in the group, the commando may be reinforced with artillery, engineers and logistical support elements. Each commando has a complement of 650 Royal Marines and Royal Navy personnel which grows to just over 700 in times of war. All three are capable of performing the full range of military operations that fall within the category of war (as in the Falklands) or operations other than war, such as their contribution to the humanitarian effort for the Kurds after the Gulf War, for which 40 Commando has, uniquely in the armed forces, been awarded three Wilkinson Swords of Peace.

Each Commando is equipped with a standard level of firepower which it is able to carry within the group. These include twenty-four Milan anti-armour missile launchers (range 1,950 metres); nine eight-millimetre mortars (range 5,650 metres); nine fifty-one-millimetre mortars (range 1,000 metres); a hundred ninety-four-millimetre antitank weapons (four per section); thirteen sustained fire (SF) machine-guns; sixteen snipers armed with L96 7.62-mm rifles; eighteen 0.5-inch Browning machine-guns; twelve 0.5-inch infantry support weapons (ISWs).

The Commando Brigade also has additional combat support units to keep an enemy occupied while commando units manoeuvre into assault positions and to amplify the combat power of the brigade. Additionally, 29 Commando Brigade, Royal Artillery, directly supports the force with three batteries of 105-millimetre light guns. Each battery has six guns and reinforces each of the commando units. A further six guns are available from a Territorial Army Commando Battery.

As amply demonstrated in these pages, the Commando ethos of carrying everything needed for at least the first 24 hours of an operation, and possibly much longer, provides 3 Commando Brigade with a unique ability to deploy at short notice anywhere in the world and mount an amphibious assault or poise offshore in a strategic demonstration of military force to deter an aggressor. This, combined with the maritime forces' facility of being able to move hundreds of miles a day over two-thirds of the world's surface, provides additional flexibility which is not available to traditional forces.

There have, however, been a number of adjustments to the commando tactics that were first displayed in the Dieppe and D-Day landings. Full-frontal amphibious assaults by storming the beaches of enemy territory are, in today's world of hi-tech defence systems, too costly both in men and shipping. Today, the emphasis is for the landing force to assault where the enemy's defences are weakest or, better still, non-existent. Current naval thinking also demands a broad space from which to strike at will, projecting all aspects of naval firepower in support of the assault on land to achieve a rapid build-up of combat power ashore before the enemy has time to react, again demonstrated by the Falklands landings, which were generally unopposed.

Because of this requirement and the need for a swift arrival ashore by the assault force, landing craft and helicopters for such operations are controlled by the Commander, Amphibious Task Force (CATF), to achieve a rapid offloading of an assault force and its heavy logistical support. A dedicated logistics unit, which will have a complete resupply chain from the initial assault through to the final phase of the operation, was long ago recognised as an essential element of commando raids but is not always effectively organised. Today, the unit is charged with providing a ready-to-move source of stores for the first 60 days of any operation, from ships to allow the brigade to operate as a totally self-sufficient force without requiring a major airlift of supplies.

As the new millennium dawned, new equipment was already sign-posted to further bolster the brigade's capabilities and performance. By the end of the last century, all its older Rigid Raiding Craft Mk2 had been replaced by the swift and powerful Mk3 version. The new raider has inboard diesel engines giving greater speed, manoeuvrability and range. Its own commando carrier, HMS *Ocean*, came into service in 1999 to increase dramatically the lift capability of the amphibious fleet with her ability to carry a commando unit with its equipment, and by 2005 it was planned to have dedicated ships available with the capacity to carry the whole fighting echelon of 3 Commando Brigade for the first time in many years. The brigade was also to receive an updated helicopter facility coming on-stream by 2006.

And, of course, all the above firepower is designed to underpin and support the skills of the men themselves, those who still wear the green beret introduced in 1940. It remains the distinctive hallmark of the Commando troops, and those who wear it have come through the gruelling tests of endurance laid on them by the various disciplines in the renowned Commando Training Course, men who have displayed qualities of individual strength as well as teamwork and 'cheerfulness in adversity, a person who will not give up and refuses to accept a situation as totally hopeless'. The basic elements of the course are surprisingly unchanged from those originally introduced at Achnacarry in 1942. There are, however, many more demanding disciplines in achieving Commando skills in a hi-tech environment. All commando training is carried out at the Commando Training Centre at Lympstone near Exeter. Royal Marine recruits complete their 30-week basic training there, the longest for infantry soldiers anywhere in the world while potential officers spend 14 months at Lympstone. Both must pass the Commando course, which is a separate four-week phase of their respective training courses, and thereafter will be entitled to wear the green beret.

There are no academic requirements for acceptance into the ranks of the Royal Marines, although recruits have to pass basic written tests covering literacy, numeracy, reasoning and mechanical aptitude. Additional training is carried out at various locations and climates, from northern Norway to the jungles of Brunei and at sea. After initial service as a rifleman, there are also training courses offering recognised qualifications in a wide range of trades and skills. Recruits also need to be physically fit enough to complete the demanding commando course. Potential officers need at least two A-levels plus three GCSEs and enter a very demanding training programme at the beginning of which they are screened three times to ensure that they have the 'physical and mental requirements to lead some of the world's most elite soldiers'.

As a crucial element of the British military's international strike force the marines and the Royal Navy provide the modern day Commando capability as part of the Amphibious Ready Group (ARG), which spends around six months of every year on exercises at sea, generally around the

Mediterranean and North Africa. The group was perfectly placed to demonstrate its 'readiness' in May 2000 when the former British colony of Sierra Leone was in danger of collapsing in turmoil. A United Nations peace-keeping force had been unable to quell the bitter, bloody conflict between government and rebel troops. As the fighting worsened, scores of British aid workers and expatriates became trapped by the fighting and more than 500 UN soldiers were taken hostage by the rebels.

Britain dispatched what its military strategists believe to be the model of modern strike-power for such situations: a three-pronged force totalling 1,600 men; the largest deployment of British troops since the Kosovo crisis in 1999. It was made up of an SAS squadron for clandestine operations and intelligence gathering, 700 lightly equipped men from the rapid response spearhead battalion of the Parachute Regiment; and the Royal Marines Commandos of the ARG. More than 800 men from 42 and 29 Commando were already aboard the Royal Navy's new helicopter assault ship *Ocean* – fully equipped with helicopters and landing craft – which was in Marseilles on an exercise when the crises blew up. *Ocean* set sail at the head of a small flotilla of three support ships and a frigate, *Chatham*. The British aircraft carrier *Illustrious* was also diverted towards Sierra Leone carrying 16 Harrier aircraft and three helicopters.

The extent of the British fire power assembled within a matter of days surprised observers and there was some apprehension about the British government's policy of putting the troops in to support the UN operation and not using them in an assault role. The paras took with them heavy and light machine-guns, mortars and Milan antitank weapons in addition to their SA-80 rifles. They travelled in two RAF Tristar aircraft, while much of their equipment, including vehicles, was delivered in a chartered Antonov 124 transport aircraft.

The marines, already on routine deployment in the Mediterranean, had a substantial inventory available: 2 Lynx helicopters equipped with antitank missiles, 6 Sea King helicopters and 4 RAF Chinook transport helicopters. Their heavy equipment included a battery of six 105-millimetre Royal Artillery guns, plus BV-206 tracked vehicles, trucks and Land-Rovers carried on the support ships, *Sir Bedevere*, *Sir Tristram* and *Fort Austin*. In the event, the British government maintained its stance of a non-assault role and the Commandos completed their mission in good order and were withdrawn on 15 June 2000. It was, however, an example of their modern capability when the need arises.

INDEX

Vessels are given in *italic.*

1 Commando Brigade 177, 179
2 Commando Brigade 180–2
3 Commando Brigade 171–4, 184–5, 226
  in Northern Ireland 238
  in Falkland Islands 242–56
  in Iraq 261–5
  recent times 257
6th Airborne Division 154, 159–61
29 Commando Regiment Royal Artillery 244, 270
59 Commando Squadron Royal Engineers 244

Absalom, Jim 185
Achnacarry Castle *see* Commando Basic Training Centre
Aden 231–4
Afrika Korps 43, 126
Air Landing Brigades
  1st 130
  6th 154
airborne troops, German 7, 14
Aldridge, Marine Jimmy 200–1, 204–5
Allen, Lieutenant A. C. 118
Ambassador, Operation 25
Antelope, HMS 246–7
Antrim, HMS 242–3
Anzio 148
Appleyard, Lieutenant Geoffrey 110–12
Ardent, HMS 245–6
Arnhem 174
Arran, Island of 32, 36
Ascension Island 242–3
Atlantic Conveyor, SS 248
Auchinleck, General Sir Claude 53–4
Avalanche, Operation 135–9
Azores 37

Bagnold, Major Arthur 44–5
Balkans *see* Yugoslavia
Bamfield, Marine 200–1, 205
Bangalore torpedoes 87, 95
Bass, USS 211
Battle of Britain 17, 28
BBC 83
Beattie, Lieutenant-Commander S. H. 69, 75
Bedfordshire Regiment 8
BEF *see* British Expeditionary Force
Berck, France 22, 24
beret, green 184
de la Billiere, Major Peter 235
Black Watch of Canada 101
Blitzkrieg 6, 14
Boer War 7–9

Boom Patrol Boat 78
Borneo 234–8
Boulogne, France 22, 24, 79–84
Bourne, Sir Alan 19, 28
brainwashing 208, 222–3
Brand, Lieutenant D. 129
Briggs, General Sir Harold 194
Brisk, Operation 37
British Expeditionary Force 1, 6–7, 14–15
Brooke, Field Marshal Sir Alan 86, 143
Brunei 234–6
Bruneval 77
Buchanan, USS *see Campbeltown,* HMS
Bulwark, HMS 232–3
Burbridge, Captain G. W. 129
Burchett, Wilfred 219
Burma 169–72
Burma Rifles 169
Burt, Gordon 228
Buzo Tactico 239–40

C-boats 22, 265
Caen Canal 155
Calvert, 'Mad Mike' 29, 145, 170, 194–5
Cameron, Captain Alastair 253
Campbell, Major Malcolm 78
Campbeltown, HMS 68–75
Canadian troops 79–84, 87, 89, 93, 103
Canberra, SS 244
canoes, submersible 78–9
Carden, Admiral Sackville 18
Caroe, Chris 250
Carr, Lieutenant 158
Cawthorne, Matthew 260–1
Chamberlain, Neville 5–6, 17
Channel Islands 24–7
  see also Sark
Chant, Lieutenant S. W. 71–2
Chappell, Ernest 11, 24, 26, 65–6, 68–73
Chariot, Operation 67–75
Chasseurs Alpins, French 8
Chindits 145, 169–70
Chitty, Provost-Sergeant Bill 40
Churchill, Brigadier Tom 134–5, 148
Churchill, Lieutenant-Colonel Jack 41, 133–4, 136–8, 148–52
Churchill, Randolph 30–1, 33, 134
Churchill, Winston 1, 17, 53–4, 259
  orders formation of Commandos 18–19
  orders formation of Parachute Regiment 36
  critical of Wavell 43
  appoints Mountbatten to Combined Operations 61–3
  after Dieppe raid 105–6
Churchill tank 89, 101
Claret operations 235–6
Claymore, Operation 37–40
Clogstoun-Wilmott, Lieutenant-Commander Nigel 46–7, 126
  see also Combined Operations Pilotage Parties
clothing 9, 11
Cockleshell Heroes 3, 84, 120–3
Collar, Operation 22
Comacchio, Lake 181–2
Combined Operations 28, 61
  Development Centre 78
  Pilotage Parties 47, 125–9, 155, 171
  see also Keyes, Admiral Sir Roger; Mountbatten, Lord Louis
Combined Training Centre, Inveraray 30
Commando Basic Training Centre 64, 231–4, 271
Commando Reconnaissance Troop 242–3
Commando School 185
Commandos
  derivation of name 7, 268
  formed 19–22

No 1: 127–8, 170, 173
No 2: 68–9, 130, 133–4, 148–52, 180–1
see also Special Air Service, 11 Battalion
No 3: 25, 35–7, 40–2, 93–4, 130–3, 145–6, 153
No 4: 35, 37–9, 79–84, 90–1, 93–9, 153, 159–60, 175–6
No 5: 35, 170, 172–3
No 6: 35, 127, 153
No 7: 35–6, 48–52
No 8: 30–3, 35–6
No 9: 148, 180–1
No 10 (Inter-Allied): 102, 147
No 11: 35–6, 48, 52–3
No 12: 40
No 30: 125
No 62: *see* Small Scale Raiding Force
Army, disbanded 183–4
Belgian 175
Middle East 37, 43–4, 50–2
Norwegian 175
Royal Marines *see* Royal Marines
Royal Navy 103, 127
Communism, rise of 185, 190–5
concentration camps 179
Condron, Andrew Melvin 209–23
Cook, George 15–16, 92, 95–7
Cooper, Group Captain T. B. 118
Cooper, Lieutenant Noel 129
Copland, Major Bill 35
Corso, Colonel Philip 223
Cosgrove, Henry 155, 158, 164–5, 177–80
Courtney, Lieutenant Roger 'Jumbo' 32, 36, 46–7, 49, 53, 125
Coventry, HMS 248
Crete 50–2
Crossley, Sub-Lieutenant A. 129
Crusader, Operation 54–8
Cyprus 225–6, 230–1

D-Day *see* Overlord, Operation
Dakota aircraft 131
Davies, Bernard 38–9, 81, 159
Davies, Major P. G. 'Pug' 236
Day, Quartermaster Sergeant 200–1, 205
De Kock, Lieutenant P. 129
Denmark 7
Dewing, Major-General R. H. 19–20
Dieppe raid 84–106
Dockerill, Sergeant 71
Douglas-Home, Sir Alec (Lord Dunglass) 28
Drysdale, Lieutenant-Colonel D. B. 199
Dunera, HMS 131
Dunkirk 1, 14–15
Durnford-Slater, Lieutenant-Colonel John 25–6, 40, 93, 133
Durrant, Sergeant Tom 73, 75

E-boats 81–2
Eden, Anthony 27, 225–6, 230
Effendi, Yassim 236
Eisenhower, President Dwight 128, 153, 230
Elverum, Norway 7
Empire Fowey, SS 230
Endurance, HMS 239, 241
Enigma encryption machine 49–50
Ensleigh, Lieutenant 192, 194
Essex Scottish Regiment 101
Eureka motor launches 69

Fairfield, Jim 239
Falkland Islands 239–56
Farrar-Hockley, Captain Anthony 205
Fearless, HMS 247
Fink, Major Gerry 202, 204
Finland 6, 8
Folboat Section 32–3
Folder, Sub-Lieutenant E. 129
Force Z *see* Layforce
Foreign Legion, French 8
Fort Austin RFA 272

Fox, Chris 251, 253–4
Frankton, Operation *see* Cockleshell Heroes
Freeberry, Corporal 57
French forces 6, 8, 14–15
  see also Vichy French
French Foreign Legion 226
Freshman, Operation 116–20
Freyberg, General Bernard 50–1
Fyson, Lieutenant Richard 181

Gale, Major-General Richard 154
Gallipoli 5, 18
Gardiner, Captain Ian 245–56
George, Lieutenant Colin 236
Gibraltar 17
Gilbert, Martin 18
Gilchrist, Donald 98
Gironde raid *see* Cockleshell Heroes
Glenearn, HMS 37
Glengyle, HMS 31–2, 36, 48–9, 52
Glenroy, HMS 36
glider operations 14, 50, 118–20, 130–1, 154
Glorious, HMS 15
Goatley boats 65, 181
Graham, Major Freddie 50, 52
Green, Jim 190–5, 226
Grivas, Colonel *see* Cyprus
Gubbins, Lieutenant-Colonel Colin 8, 13
Guernsey *see* Channel Islands
guerrilla warfare 8
Gulf War 260–1
gunboats, steam 94
Gurkha Rifles 169–70, 235, 244
Gustav Line 147–8

Halifax bombers 117–20
Hamilton, Nigel 104
Hardelot, France 24
Harden, Lance-Corporal H. E. 177
Harding, General Sir John 194
Hart, Lieutenant A. 129
Haselden, Captain Jock 55
Hasler, Major H. G. 'Blondie' 78, 84, 121–2, 125, 171
Hastings aircraft 228
Haydon, Brigadier Charles 59, 64
Hayes, Graham 111–12
Head, Alan 87
Healey, Denis 241
Hefferson, Jim 102
Henderson, Lieutenant Ian 71
Henriques, Colonel Robert 83
Herefordshire Regiment 10
Hertfordshire Regiment 8
Hicks, Marine 200–1
Hill, Brigadier James 89–90, 154
Hitler, Adolf 6–7, 16, 107
  and 'Commando Order' 1–3, 115–16
Holden-White, Major Harry 83
Holland, Major J. C. F. 8
Hopkins, Lieutenant 71
Horsa gliders 118–20, 131
Houghton, Major Titch 102
Hugh-Hallett, Captain J. 116
Hunt, Sir Rex 239–41
Hunter, Corporal Tom 181

Illustrious HMS 272
Imperial War Museum 80–1
In Which We Serve (film) 61–2
Independent companies 9–16, 21–2, 25, 27, 35
Ingles, Lieutenant 54
Ingleson, Lance-Bombardier 256
interrogation, resistance to 208
Intrepid, HMS 247
Inverailort Castle 29
Iraq 231–2, 261–5
Ismay, General Hastings 19, 24
Italian navy 67, 78

Jellicoe, Lieutenant Lord George 59
Jersey *see* Channel Islands
Jones, Lieutenant-Colonel 'H' 244
Jordan 231
Jubilee, Operation 84–106

Junkers Ju-52 7, 14, 50
Junkers Ju-87 14, 50–1

Kangaw, battle of 173
Keeling, Major-General A. M. 236
Kelly, HMS 61
Kemp, Peter 116
Keyes, Admiral Sir Roger 28–9, 54–6, 59, 61
Keyes, Lieutenant-Colonel Geoffrey 54–5
King's Liverpool Regiment 169
Kiruna, Sweden 6
Kitchener, General Horatio 18
Knowland, Lieutenant G. A. 173
Korean War 197
Kurds, Iraqi 261–5
Kuwait 232

Lancaster bomber 178
Landing Craft 80, 94
Langton, Lieutenant Tommy 56–7
Lassen, Lieutenant Lars 113
Lassen, Major Anders 181–2
Lawrence, T. E. (of Arabia) 8
Laycock, Brigadier Bob 32, 43, 50–2, 54–5, 59, 130, 134–5
Layforce 37–8, 43, 45, 47–9, 52–3, 58
Le Touquet 23
Lee-Enfield rifle 12
Leicester, Brigadier 'Jumbo' 22, 153, 162
Lentini Bridge 132–3
Levick, Commander Murray 30
levies, Arab 8
Lewes, Lieutenant Jock 53
Litani River, Lebanon 52
Liverpool, HMS 210
Lloyd-George, David 18
Lofoten Islands 37–40, 114, 153
Long Range Desert Group (LRDG) 44–5, 53–4, 126
Lovat, Brigadier Lord 27, 29, 37–9, 42, 79–83, 91–3, 153
Luftwaffe 18, 41, 43, 100–1, 103
Lumsden, Lieutenant-Colonel B. J. D. 137

MacKay, Sergeant Ian 244–5
Mackesy, Major-General P. J. 8
Madden Committee 5, 7, 143
Madoc, Major-General R. W. 227–9
Maginot Line 6
Maid Honor Force 110–12
Makarios, Archbishop 231
  see also Cyprus
Malaya 190–5
Malayan Scouts 195
Malta 17
Manners, Lieutenant-Colonel 'Pops' 146, 151
March-Phillips, Gus 110–12
Mayne, Captain Paddy 53, 130–3
McCool, Brian 85
Messenger, Charles 49, 80
Methuen, Lieutenant D. A. 118
Milch, Field Marshal Erhard 177, 179
Military Intelligence (Research) 8–9
Millin, Piper Bill 159
Mills, Lieutenant Keith 240
Mills-Roberts, Brigadier Derek 161–2, 177, 179
Mo, Norway 12
Montanaro, Captain Gerald 79
Montgomery, Bob 67
Montgomery, Field Marshal Bernard 86–7, 104
Montgomery, Lance-Corporal 254
Morris, Lieutenant A. D. M. 236
Mosjøen, Norway 13–14
Moss, RSM 71
Motorman, Operation 238
Mountbatten, Lord Louis 61–8, 77–8, 83–5, 92, 103–4, 117–18
  in Far East 145, 167–74
Musketeer, Operation 226–30

Namsos, Norway 8–9
Narvik, Norway 6–9, 12, 15
Nasser, President 225
NATO 257–8, 265
Neillands, Robin 152
Newman, Lieutenant-Colonel Charles 69, 75
Nicholson, Captain Godfrey 30, 32
Noott, Major Gary 239
Norman, Major Mike 239–41
Norris, Vice-Admiral Sir Charles 77
Northern Ireland 238, 257, 265–7
Norway 4, 6–8, 116–20
  Independent company operations in 12–14
  Operation Claymore 37–40
Nott, John 241
Nunn, Captain 242

Ocean, HMS 271, 272
Offensive Operations 19, 28
  see also Combined Operations
Oran landings 83
Oslo, Norway 7, 9, 15
Osnabrück 179
Overlord, Operation 141–2, 153–65

Palestine 52–3, 186–90
Pantelleria Island 37
Parachute Brigades 154
parachute operations 14, 154
Parachute Regiment 36, 77, 89, 226–8, 235, 244, 272
Paterson, Major J. R. 13
Paul, Leutnant-Kapitän Fritz 74
Pearl Harbor 1
Pegasus Bridge 155
Peskett, John 199–208
Pétain, Marshal Henri 16
Pettersen, Lieutenant-Colonel Sverre 118
'Phoney War' 6
Picton-Philips, Lieutenant-Colonel 102
Pike, General Hew 245
Pinkney, Captain Philip 113
Pittard, Frances 113, 115
Plymouth, HMS 242
Poland, invasion of 5–6, 8
Polish forces 8
Porteous, Colonel Patrick 91, 93–5, 97–100
Pound, Sir Dudley 62
Preece, Trooper Freddie 79
Prince Albert, HMS 91–2, 132
Princess Beatrix, HMS 37
prisoners, treatment of 107–10, 114–16, 119–20, 200–8, 213–23
Pritchard, Bill 67

QEII, SS 258
Queen Elizabeth, HMS 78
Queen Emma, HMS 37–8, 133

Radfan *see* Aden
Rangers, American 89, 103, 148
Rhodes 36–7, 46–7
Richards, George 187–9
Rimau, Operation 78n
Rommel, General Erwin 43, 54–5, 58, 84
Roosevelt, President Franklin D. 86
Royal Air Force (RAF) 17–18, 36, 86, 103, 172–3, 175
Royal Army Service Corps (RASC) 9
Royal Canadian Air Force 103
Royal Engineers, Corps of (RE) 9, 79, 118
Royal Marines
  40 Commando 102, 130, 145–6, 180–1, 227–8
    in Borneo 236
    in Northern Ireland 266–7
  41 Commando 125, 130, 137–8, 153, 162–3
  41 Independent Commando 199, 211
  42 Commando 170, 187, 192, 227–9
    in Borneo 236–7

43 Commando 148, 180–1
44 Commando 170, 172
45 Commando 155, 177, 227–9, 231–2, 234
46 Commando 153, 155, 164, 175
47 Commando 153, 155, 176
48 Commando 153, 163–4, 176
Boom Patrol Detachment 78, 121, 125
on Dieppe raid 89, 93
today 269–71
see also 3 Commando Brigade
Royal Naval Volunteer Reserve (RNVR) 25
Royal Navy (RN) 8–9, 15, 52–3, 82
Commandos 103
Royal Oak, HMS 38
Royal Tank Regiment, 7th 51
Royal Welch Fusiliers 11
Ryder, Commander R. E. D. 69, 75

Saddam Hussein 232, 261–2, 265
Salerno 135–9
Santa Fe (submarine) 243
Sark 113–15
see also Channel Islands
Savage, Able Seaman 75
Scots Guards, 5th Battalion 6
Scott, Lieutenant-Commander Peter 94–5
Sea Reconnaissance Units (SRUs) 171
Selborne, Lord 120–3
Severn, Corporal 54
Sheridan, Major Guy 242–3
Sherwood, James 30–2, 45–8
Sicily 130–5
Siddons, Major 10–11
Sierra Leone 271–2
Singapore 78, 84
Sir Bedevere, RFS 272
Sir Galahad, RFA 244, 248
Sir Tristram, RFA 244, 248, 272
ski manoeuvres 6, 8
Sledgehammer, Operation 84
Sleeping Beauty *see* canoes, submersible
Slim, Field Marshal William 168, 170, 172
Small Operations Group (SOG) 171
Small Scale Raiding Force 110–16, 125
Smith, Lieutenant Bob 129
South Georgia 240, 242–3
Southby-Tailyour, Major Ewen 242
Soviet–Finnish dispute 6, 8
Sparks, Ned 122
Spearman, William 80–3, 108–10, 156
Special Air Service (SAS) 29, 45, 53, 113, 232–3, 242–3, 272
11 Battalion 36
Special Boat Service (SBS) 32, 36, 55, 59, 171, 243, 258
Special Boat Squadron (SAS) *see* Special Boat Service (SBS)
Special Brigade 35
Special Operations Executive (SOE) 110–11
Special Raiding Squadron *see* Mayne, Captain Paddy
Special Service Battalions 35, 153, 162, 170, 175, 177
St-Nazaire 66–75
Stalin, Joseph 6
Stanley, Lieutenant-Commander 243
Stevens, Lieutenant-Colonel T. M. P. 233
Stewart, David 251
Stilwell, General Joe 170
Stirling, David 29, 45, 53–4, 58
see also Special Air Service (SAS)
Stirling, Major Bill 29, 116
Sturges, Major-General Robert 144, 152
submarines 18, 54–5
Suez 37, 225–30
Sukarno, Doctor 234–5
Sunfish, HMS 38

Swayne, Sir Ronald 9–10, 12, 14, 21–7, 29–30, 35–6, 74
Sweden 6

Talisman, HMS 55–6
Tanganyika 233
Tasker, Peter 162
Teacher, Lieutenant-Commander Norman 129
Terry, Sergeant Jack 55–7
Thatcher, Margaret 239, 241–3
The Exodus (US ship) 187
Tidespring, RFA 242–3
Tirpitz (battleship) 67
Tito, Marshal (Josip Broz) 148–9, 180
Tod, Brigadier Ronnie 21–4, 180
Tollemache, Colonel Humphrey 171
Torbay, HMS 54–5
Torch, Operation 59, 83–4, 126
training 64–6, 91–2, 142, 185, 271
transport 9
Triumph, HMS 46–7, 49
Trondheim, Norway 8, 15

Ulster Monarch, HMS 131
Ulster Prince, HMS 13
United Nations (UN) 187–8, 197, 229–30, 235, 261, 265, 272
US Divisions 154

Vaagso raid 40–2, 64
Valetta aircraft 228
Valiant, HMS 78
Vaughan, Lieutenant-Colonel Charles 64
Vichy French 16, 48, 52
Volunteer Reserve, Royal Marine Forces 185

Walcheren operation 175–6
Walker, General Sir Walter 169, 235
Warsash (RN training location) 30
Warspite, HMS 9
Washbrook, Captain 205
Watson, Marine 254
Waugh, Evelyn 10, 50, 52
Wavell, General Sir Archibald 43–5, 48, 52
Wellington, Captain the Duke of 138–40
Wesel (Rhine) 178
Wessex helicopter 243
Weston, General 52
White, Knocker 152
de Wiart, General Carton 8–9
Wilson, Harold 234
Wingate, Major-General Orde 29, 145, 168–70
Winnington. Alan 218–19
Workshop, Operation 37

X Troop 102

Young, Lieutenant-Colonel Peter 146, 173
Yugoslavia 148–52, 180, 265

Zeebrugge raids (1918) 28
Ziegler, Philip 61, 63, 67, 105